SUN POWER

ALSO BY NEVILLE WILLIAMS

Chasing the Sun: Solar Adventures Around the World

SUN POWER

❋ How Energy from the Sun Is Changing Lives ❋
Around the World, Empowering America,
and Saving the Planet

NEVILLE WILLIAMS

A TOM DOHERTY ASSOCIATES BOOK ■ NEW YORK

SUN POWER: HOW ENERGY FROM THE SUN IS CHANGING LIVES AROUND THE
WORLD, EMPOWERING AMERICA, AND SAVING THE PLANET

A Forge Book
Published by Tom Doherty Associates, LLC
175 Fifth Avenue
New York, NY 10010

www.tor-forge.com

Forge® is a registered trademark of Tom Doherty Associates, LLC.

Library of Congress Cataloging-in-Publication Data

Williams, Neville, 1943–
    Sun power : how energy from the sun is changing lives around the world,
empowering America, and saving the planet / Neville Williams. — First edition.
        p. cm.
    "A Tom Doherty Associates Book."
    ISBN 978-0-7653-3377-3 (hardcover)
    ISBN 978-1-4668-0540-8 (e-book)
    1. Williams, Neville, 1943–   2. Solar energy industries.   3. Solar
energy industries—Developing countries.   4. Solar energy industries—
United States.   5. Photovoltaic power generation—Developing countries.
6. Photovoltaic power generation—United States.   7. Solar energy—
Government policy—United States.   I. Title.   II. Title: How energy from
the sun is changing lives around the world, empowering America, and
saving the planet.
HD9681.A2W55 2013
333.792'3—dc23

                                                                    2013025795

Forge books may be purchased for educational, business, or promotional use. For information on bulk purchases, please contact Macmillan Corporate and Premium Sales Department at 1-800-221-7945, extension 5442, or write specialmarkets@macmillan.com.

First Edition: April 2014

Printed in the United States of America

0 9 8 7 6 5 4 3 2 1

Dedicated to the memory of the late Hermann Scheer,
the world's greatest solar visionary

CONTENTS

Global civilization can only escape the life-threatening fossil-fuel resource trap if every effort is made to bring about an immediate transition to renewable and environmentally sustainable resources and thereby end the dependence on fossil fuels. Making the groundbreaking transition to an economy based on solar energy and solar resources will do more to safeguard our future than any other economic development since the Industrial Revolution.

—HERMANN SCHEER
The Solar Economy (*Solare Weltwirtschaft*, 1999)

The most important thing in the world is to have the love of God in your heart. The next most important thing is to have electricity in your house.

—TENNESSEE FARMER, 1940
(inscribed over the entrance to the electricity exhibit at the Henry Ford Museum, Dearborn, Michigan)

I'd put my money on solar energy. What a source of power! I hope we don't have to wait until oil and coal run out before we tackle that. I wish I had more years left.

—THOMAS ALVA EDISON

Solar power provides a marvelous alternative source of energy, particularly in remote places. It has been proven in the harshest of environments both terrestrial and extraterrestrial. With solar power, the village in the jungle can leapfrog into the modern age.

—Sir Arthur C. Clarke

America used to be a country that thought big about the future. Major public projects, from the Erie Canal to the interstate highway system, used to be a well-understood component of our national greatness. Nowadays, however, the only big projects politicians are willing to undertake—with expense no object—seem to be wars.

—Paul Krugman
The New York Times, April 13, 2012

PROLOGUE

In 2010, my old friend Tom Tatum called me with a proposal for a large solar project. I thought I was done with solar, but he had other ideas. Tom got me into solar at the U.S. Department of Energy back in 1979. He was working as President Carter's solar energy liaison with Capitol Hill and federal agencies. We stayed in touch ever since, despite going separate professional and geographic ways. Tom has a ranch in northern New Mexico, fourteen hundred acres of dry sage on the slopes of ten-thousand-foot Ute Peak that overlook the fertile San Luis Valley. A one-kilowatt solar-powered deep-well pump provides water to guest cattle and large herds of elk that share the meager grazing.

"Let's do a solar project on my ranch," he said.

"Who will buy the power?" I replied.

"I'll talk to Kit Carson Co-op in Taos," he said. And he did.

Luis Reyes, CEO of the Kit Carson Electric Cooperative, one of the most progressive and visionary electric cooperative managers in the nation, was interested. I flew out to walk the ranch and to meet with Tom and the Kit Carson engineers. There was a perfect section of one hundred acres on the ranch that could accommodate 10 megawatts of solar. I contacted Suntech, the world's largest solar company at the time, and they said they could finance and build it; they sent out one of their project executives to see the site and meet with Tom and me.

Then, we discovered that the ranch was a little too far from the nearest substation and that new high-voltage lines would be required. This could have been overcome, but the bigger problem was that Tri-State Energy, which provides wholesale coal-fired power to its member co-ops through long-term contracts, would not permit members to buy more than 5 percent of their electricity—in this case sun power—out of contract. This bedeviled Luis Reyes, who fights for more solar at every turn and may one day get Tri-State to reconsider its policy. Luis believes, "Part of our culture in New Mexico is sustainability and so we want to be a renewable, green community. We want to make solar the rule, not the exception." We were all angered that coal-generating plant owners and operators were able to block sun power from providing as much electricity as Kit Carson wanted to purchase. We retrenched, but didn't give up.

We wanted an open energy market, not protected monopolies, so the fight was on. We were mad, and determined to take on the coal boys any way we could. It was clear to us right then that the future of electricity in this country was a battle between coal and renewable energy, including solar. We believed solar would win . . . one day.

Luis agreed, and he suggested we build a smaller solar power plant on a piece of land he knew about close to a substation just two miles south of the Colorado border, at Amalia. It was owned by the Rio Costilla Cooperative Livestock Association, at Costilla, just up the road from Tom's ranch house. The cattlemen liked the idea of leasing fifteen acres for a solar array. Luis said he'd buy 1.5 megawatts of direct current (DC) of solar, but Suntech said they weren't interested in building anything under 10 megawatts.

So I took the project to my former company, Standard Solar, which you will read about in chapter 11, and they said they'd love to do it. They agreed to finance and maybe own part of it. When Luis came to Washington, D.C., for a meeting of the National Rural Electric Cooperative Association (he's also on the board of the Solar Electric Power Association) I set up a breakfast at the Grand Hyatt Washington with myself and Standard Solar's CEO, Tony Clifford. Tony and I wore suits. Luis arrived New Mexico style, in a polo shirt. He charmed us with his eagerness to "go solar" and told us how he would "socialize" the additional cost of sun power across the rate base. New Mexico required utilities to purchase a set amount of solar electricity, but that wasn't what was driving Luis. He knew that solar was the future, that New Mexico had more sun than almost anywhere else in the continental United States, and that the cost of coal-generated power would rise, whereas solar purchased on a multi-year contract would be fixed, even if it cost more initially. And he wanted to do the right thing, and so did most New Mexicans.

All that remained were a few minor details like acquiring a lease on the land, getting county planning and zoning approval, and having the project OK'd by the Kit Carson board based on a price for the solar kilowatts to be negotiated. Tom put on his cowboy hat and boots and began attending meetings with Luis and his staff, with the county authorities, with lawyers, and with the media, which supported the project from the start. Tony flew out from D.C. to negotiate the price for the power. We negotiated a twenty-five-year land lease with the cattlemen's cooperative. We soon got unanimous approval from the county zoning board. Even with cooperation and goodwill on all fronts, the entire development process took two years! In America today, nothing is done with alacrity.

The deal finally came together in August 2011, and we broke ground on beautiful Wild Horse Mesa, at eight thousand feet above sea level. Standard's subcontractor, Paradise Power Company of Taos, worked through the winter of 2011–12 as best it could. Northern Taos County was thrilled to have the jobs. Standard purchased container loads of 280-Wp polycrystalline panels from China's Trina Solar, which were shipped to the site (see the glossary for an explanation of technical terms and abbreviations). Washington Gas Energy Services, from D.C., came in to finance and take an equity stake in the project.

A note here on Paradise Power Company of Taos. This is the second-oldest family-owned solar company in the United States, started by Michael Weinman in 1979. An electrical engineer living off the grid in remote northern New Mexico, he relied on kerosene for light. One day he bought two thirty-watt ARCO Solar modules that radically changed his life. He hooked them up to batteries and soon was living like the families in the developing world you will be reading about in the following chapters, with bright fluorescent lights in every room. He began supplying "hippie homes" with solarelectric

Wide view of small part of 1.5-megawatt power plant on Wild Horse Mesa, northern New Mexico, sending power to Kit Carson Electric Co-op. *(Author)*

systems he put together himself while teaching his young son, Dan, how the technology worked and how to run a business.

Like John Schaeffer's Real Goods, the oldest existing solar supplier, which initially provided California's marijuana growers with off-grid battery-powered solar systems in the seventies, Michael's company sold sun power to the growing counterculture community of Taos County, many of whom arrived on the heels of then Taos resident Dennis Hopper and his film *Easy Rider* (1969). They're still here, and in 2009 they celebrated the fortieth anniversary of the "Taos Summer of Love." Today, many live in solar-powered eco-homes in the Earthship Community subdivision. Paradise Power's first big job was providing solar power to the local radio station, KTAO, with a large photovoltaic array. Today, the town hosts the annual Taos Solar Music Festival.

Paradise Power is now run by Michael's son, Dan, who started

Tom Tatum speaking at 2012 inauguration of Amalia solar plant, New Mexico. *(Standard Solar)*

working for his father when he was eight years old. The company installs grid-tied residential solar home systems and constructs large-scale solarelectric projects such as the 500-kilowatt solar system on the Taos campus of the University of New Mexico. They also roofed the Kit Carson Cooperative headquarters' parking lot with a huge solar shade structure. These were the right guys to build our project, which would be designed, engineered, and financed by Standard Solar. On April 10, 2012, the partially completed plant was dedicated at an outdoor ceremony attended by U.S. congressman Ben Ray Luján, state senators and representatives, and executives from Standard Solar and Washington Gas, followed by a barbecue for two hundred guests.

Tom, in his brown leather jacket and Stetson hat, addressed the gathering. He recalled his early involvement with President Carter's solar energy plan and cited President Obama's support for solar energy thirty-plus years later; the Obama administration provided stimulus funding and federal tax credits to help bring such projects as ours to fruition. "Every project has a DNA. This project combines federal government, state, electric cooperatives, county, and a rural community. Today, we are dedicating this project that will supply the daytime electricity needs of the farmers, ranchers, and communities in northern Taos County for the next twenty-five years."

The project went "on-line" in June 2012, and Standard Solar's Mike Brown, who supervised the project through the winter from his trailer at the remote site, was able to go back to Maryland. We'd hired Mike at Standard Solar in 2008, an out-of-work marine carpenter keen to get into the solar business, but with no experience. Now, he was a construction manager on utility-scale solar projects, loving it, and well loved by all.

I toured the project in July 2012, while writing this book. Tom and I and our wives walked the rows of single-axis tracking arrays that follow the sun from east to west—eight small electric motors turn the 5,280 panels. I couldn't help but reflect on my long solar odyssey, from Tom's office at the DOE in 1979 to installing 20-watt solar panels on farmers' homes in western China to profitably selling 40-watt solar lighting systems to poor people in India to Standard Solar's four-kilowatt residential installations in suburban Washington—which were two hundred times bigger than what we'd provided to Chinese peasants.

I recalled our first 20-kilowatt system for the Potomac Electric Power Company (PEPCO) in D.C., when we couldn't believe we were doing something so big, and for a utility! Standard Solar later began selling and installing 100-, 500-, and 800-kilowatt solar systems, to my astonishment. Now, here we were, walking through fourteen acres covered by a megawatt and a half of cobalt-blue polycrystalline solar panels sparkling in the brilliant New Mexico sun, a sight I never thought I'd see.

(In 2014, we completed a second 1.7-megawatt solar plant serving the Mora San Miguel Electric Cooperative in extremely rural New Mexico near Las Vegas. We managed to beat the cost of coal-fired power purchased by the co-op by a fraction of a cent per kilowatt hour.)

Here's the kicker, and the good news. In today's solar energy marketplace *our New Mexico solar project is so small that it barely warranted mention in the trade press*! Even had we managed to build our envisioned 10-megawatt project on Tom's ranch, it would have been dwarfed by the 30-megawatt solar photovoltaic (PV) installation at Cimarron, just over the Sangre de Cristo range on Ted Turner's ranch. Our project was tiny compared to the one hundred megawatts of solar arrays covering thousands of acres in the northern end of the San Luis Valley in Colorado. Owned by Met-Life and Goldman Sachs, all of this power was going into the Colorado energy grid. (The San Luis Valley, at 7,500 feet, is ideal for solar; it's cooler than desert sites, which is good for PV, and the "insolation" is more intense.)

Then, there are the massive multimegawatt solar plants already built or coming on-line in California, Texas, and Nevada, plus the many other utility-scale PV installations in America that you will read about later in this book. To encourage more solar on the open ranges and deserts in the West, in 2012 Interior Secretary Ken Salazar signed off on seventeen new "solar energy zones" on 285,000 acres of public lands in six western states.

On top of this is the biggest solar electric business of all: *residential and commercial rooftops across the country*. This power is produced by people themselves and is known as "distributed generation," tens of thousands of dispersed solar installations connected to the power grid.

Meanwhile, I couldn't be happier that the largest solar project I could have ever envisioned being part of was just a tiny shining example of what is actually happening in the exploding world of sun power. I'd been part of that world in one way or another since working for the Carter administration so long ago, when I was still a freelance journalist. After giving up the life of a starving writer and launching a midlife career in solar energy, advocating, promoting, distributing, and selling solar energy around the world, I decided to tell this story, which will take you to many countries before bringing it all back home to America and the long-awaited solar revolution.

1

SOLAR REVOLUTION? 1979

"Why aren't we using solar energy here in America?" I have been asked whenever I've spoken about solar electricity in the developing world, where it is the only source of power for millions of people with no access to electricity. The answer I gave from 1990 to 2005 was simple: "Because we already *have* electricity and solar is still expensive."

That answer is different today. We *are* using solar, as this book explains, but not as much as we could be. Whether you know it or not, we will have a solar-powered future, if you include all renewable energy sources that claim the sun as their progenitor. Everyone knows the sun will be with us for a very long time, while fossil fuels are on the way out, so we will use solar energy because we, as planetary citizens, have no choice.

It has been forty years since Germany's rocket engineer and America's space pioneer Wernher von Braun said in Paris at the world's first global solar conference, titled The Sun in the Service of Mankind, "I believe we are at the dawn of a new age, one which might be called 'the solar age.'"

This is a book for doers and dreamers. It is a personal narrative about real people doing real things, not a science text, a policy treatise, or an academic analysis of energy, technology, or political issues. It is not a technical or "how to" handbook. I'm not going to pronounce on what we "must do" or "should do" but will explain what people *are doing* around the world and in the United States. You will meet some of the people among the millions who have chosen to put the sun at their service, from those in villages in Africa to suburban America, from rural China to Google's campus, from thousands of industrial rooftops to America's huge solar farms.

Sun Power is a practical manifesto on how we can, and are, changing the world with solar photovoltaics, an unappealing word for an elegant technology. Solar photovoltaics (PV) has been handicapped by this name from the beginning. I prefer "solarelectric" (like "hydroelectric," one word) to describe solar power generated by photovoltaic technology that converts sunlight directly into electricity. "Solar power" is the common term today, "PV" for short. I once heard a government official at the dedication of the University of Maryland's solar-powered race car call it "photogalactic." Maybe it's more galactic than voltaic, since the technology does seem out of this world.

Solar energy, the world's fastest-growing industry, is generally a boring subject. But getting and using free energy from the sun is exciting and, if humanity is to survive, ultimately necessary. The first part of the book is about bringing solar power to people who have no access to electricity. The second part is about how America is using solar power here at home and how we are able to save money while saving the planet and how America is catching up with Europe, where the solar revolution took off at the dawn of the new millennium.

Sun Power will introduce you to people who never benefited from the modern age, who never had electricity, but who have been the solar power pioneers who were the first to light their homes with the technology that we will all be using in the future. You will learn about a nonprofit promoter of solar in more than a dozen developing countries and about its spin-off, a commercial venture that has sold and installed over 150,000 solarelectric systems in India. Finally, the book tells the story of my last solar adventure as a "serial entrepreneur" and how hundreds of American entrepreneurial start-ups are "saving the world one rooftop at a time." And one "solar farm," "solar garden," and multimegawatt solar power plant at a time. You will read about the enormous opportunities in solar and how dozens of global corporations have entered the clean energy business.

The book reflects the need to focus on hope, a commodity that seems in increasingly short supply these days as we are faced with "resource wars" and the "clash of civilizations." The lethal mix of oil, Islam, and Israel renders the world a more unstable and frightening place than it has ever been. The hope is found in the delight expressed in a child's eyes as she flips a light switch in her family's wattle-and-daub home and, for the first time, watches an electric light come on. The hope is represented by the fact that this family was able to purchase their solarelectric installation on credit and that for the first time ever their world did not go dark at 6:00 P.M., as it does year-round in the equatorial latitudes. The hope is in their empowerment, and ours.

Clean energy solutions are already working in a bigger way for the whole of humanity than I could ever have imagined ten years ago. While the West fights its wars over oil, humble farmers in developing countries, along with American and European homeowners and thousands of companies and hundreds of utilities worldwide, are putting the solar solution to work. There is hope in knowing that the solar power business is growing at 80 percent a year, providing untold opportunities for a new generation of entrepreneurs.

You don't need to know or care about how this technology works, any more than you need to understand the principles of a liquid crystal display or LED screen to turn on and watch a television. You needn't be able to tell photons from electrons to appreciate the attraction of solar energy. It does not take technical knowledge to be a sun worshipper. You do need to know, if you haven't heard already, that enough energy from the sun falls on the earth in fifteen minutes to power the world for a year.

For the history of solar power and a layman's discussion of technology, I refer the reader to Appendix 1, "Solar Tech Simplified."

A portion of this book first appeared in 2005 under the title *Chasing the Sun: Solar Adventures Around the World.* It told the story of how a small group of "unrealistic" and perhaps "unreasonable" activists and entrepreneurs turned their vision into the reality of a half million people getting their household energy from the sun. At the time it begged the question, "If poor people in the developing world can afford to use solar electricity, why can't we?" That is our challenge.

* * *

It seems like a dream now, the halcyon days of the 1950s, when I was growing up during a time of innocence in a world full of promise, without cynicism or irony. The bright future lay before me, around the corner, capturing my imagination. There were no economic worries in the air of postwar America, no shortages of anything. To a high school student in a small Ohio town, whose father's metallurgical engineering job paid for a large white house on three acres, a new '57 Chevy, and a second car for his mother, who didn't need to work, it was the best of times.

I was born in 1943, at the beginning of the baby boom, when hundreds of thousands of soldiers left their wives (soon-to-be mothers) behind as they shipped off to war. The cultural icons of the boomer generation—among them, Bob Dylan (born 1941), Paul McCartney (1942), Mick Jagger (1943), and Jim Morrison (1943)—defined the zeitgeist as one of fun fun fun, wild rebellion, and indignant protest. None of us was to have a normal life.

It wasn't until I was in high school that I learned we'd won World War II by dropping two nuclear bombs on Japanese industrial cities, obliterating them and some two hundred thousand of their inhabitants. By this time, Russia had the bomb as well. I still remember bicycling six miles to town to buy an issue of *Mad* magazine, the cover of which showed planet Earth with big chunks blown out of it as Russia and the United States lobbed ever-larger intercontinental ballistic missiles back and forth at each other. Ha ha.

However, the threat of nuclear missiles was offset by President Dwight D. Eisenhower's Atoms for Peace program, which pumped billions of government dollars into building the first nuclear power plants. Soon, it was said, we would have power "too cheap to meter." You just needed a little uranium 235 and some plutonium and you'd have all the power you'd ever need, for ever and ever.

This seemed too good to be true, even to a tenth grader, so I wrote to the Atoms for Peace program in Washington, and soon a large package of technical materials arrived in our mailbox. I studied them carefully. My father helped me with the technical terms and descriptions, and soon I understood how nuclear energy was produced. Amazing! Neat!

What could be better than this?

I decided to produce a detailed, comprehensive, schematic drawing of exactly how a nuclear power plant worked for my high school science project. I mounted the drawing on a big wooden board, carted it off to school, and explained to the science teacher all about critical mass and controlled reaction and heat transfer to make steam for generators, ultimately producing electricity.

It was many years before I learned that it would be better to use photons from the sun to produce electrical energy. It would be every bit as clean as nuclear power, without the risk of contaminating our cities for twenty-five thousand years if something went wrong (and something will always go wrong, some day); it would eventually prove to be cheaper than nuclear energy, which became so costly that it nearly bankrupted several U.S. utilities that invested too heavily in it. Later, Three Mile Island, Chernobyl, and Fukushima nailed the nuclear coffin shut, perhaps for good. That leaves the sun as the natural source for electricity.

Despite this early interest in electricity production, I did not get a job in the power industry, nor did I study science, engineering, technology, or business. None of my father's astonishing abilities in mechanics, science, and engineering rubbed off on me. I decided to become a writer and a journalist; studied history, English, and polisci; and set out after college to cover the story of my generation: Vietnam. I decided to "learn by doing." I learned to write about war by going out to see it and watching contemporaries die in it, including one of my best high school friends, drafted into the infantry right after he returned from serving his country for two years in the Peace Corps. (I too was drafted, but that's not a subject for this book.)

The Vietnam War alienated a generation, and many of us "dropped out" of mainstream society. I chose the freelance writer's life in a Colorado mountain town, a place to recover from the sixties and try to become whole again. Meanwhile, the Middle East soon replaced Vietnam as a focus of our national attention. Today, it is all we think about. Radical Islam, Muslim Iran, and the Arab revolt seem to control our lives and our future.

After the 1973 oil embargo caused by the Middle Eastern members of the Organization of Petroleum Exporting Countries (OPEC), *energy* became dinner-table conversation. Soon thereafter—in the wake of the Watergate scandal, Nixon's resignation, and the United States's hightailing it out of Vietnam at long last, losing our first war—the nuclear engineer and submarine officer (and Georgia governor) Jimmy Carter was elected to preside over a very depressed United States, wearing his cardigan and telling us to turn down our thermostats to save energy.

President Carter created a new government agency, the U.S. Department of Energy (DOE), and a new cabinet post. The DOE included a division dealing exclusively with energy conservation and solar energy.

Tom Tatum was one of Carter's bright young campaign operatives, a Georgia native and Vanderbilt University law graduate who earned his political bona fides managing Maynard Jackson's campaign to become the first black mayor of Atlanta. Carter wanted to keep a close watch on his new pet projects, the DOE and his energy policy, so he brought Tom to Washington and offered him a special assignment at the DOE's Office of Energy Conservation and Solar Energy (which was soon known as Conservation and Solar).

I had first met Tom in the aforementioned Colorado mountain town, where he had recently bought a ski condo. A former antiwar activist and mutual friend, Sam Brown, who was then working for Carter in Washington, had told Tom to look me up. We met at the local saloon, where we talked serious politics amid the cowboys and ski bums, trying to hear each other over Willie Nelson on the jukebox. Scotch, our drug of choice, lubricated our late-night conversations about . . . energy! Tom said he needed help in Washington at the new Department of Energy to launch "the solar revolution."

The what?

I was rested from my antiwar activism and ready for a good fight, and a revolution sounded perfect. So, when Tom, back in Washington, called to ask me to come to D.C. immediately to help him promote solar and energy conservation at the highest levels of the U.S. government, I said sure. I came down from the mountain, spent part of 1979 in Washington at the DOE, and returned to D.C. for the summer of 1980, still one of the hottest on record. Temperatures hovered at 86 degrees *indoors* because President Carter had ordered all the government thermostats set at 80 to save energy. And you couldn't open the windows.

Despite the heat, we had not heard of global warming. That would come later. What we were concerned about, talking late into the night over our Dewar's at the Hawk and Dove on Capitol Hill, was *energy security*. Carter had come to power just after the United States had passed "peak oil," the point at which the country began to extract its oil reserves faster than it could find new oil deposits. The United States was no longer self-sufficient in the oil department, and we began to import more and more crude oil—over 50 percent of our daily needs—from the OPEC countries that had caused the long gas lines in 1973 and who were now (1979) raising prices to over thirty dollars a barrel. In July 1979, President Carter addressed the country and said, "Energy will be the immediate test of our ability to unite the nation."

The issue quickly became: How could the United States wean itself off dependence on foreign oil? Especially oil from the unstable Middle East?

Tom and I thought we knew. We read S. David Freeman's *Energy: The New Era*. He was a guru to us. Carter had appointed Freeman to head the Tennessee Valley Authority, where Freeman was closing down three nuclear reactors while preaching energy independence, energy efficiency, and solar energy. (David and I later crossed paths in a remarkable if star-crossed way, as I will explain later.) Denis Hayes, the founder (with Senator Gaylord Nelson) of Earth Day in 1970, was writing about "renewable energy" (a new phrase for me)— and the brilliant young Amory Lovins was agitating for a "soft energy path" and for "energy conservation," making his views known at our office in the DOE. Tom directly advised the president about these new policy ideas.

I moved into my office next to the assistant secretary for solar energy and energy conservation at DOE and began writing speeches and developing policy briefs. I had a U.S.-government-issue manual typewriter, since I never could manage to work an IBM Selectric. Computers? Please. I'd only seen one, and it took up about forty square feet. And I knew absolutely nothing about energy conservation or solar energy.

However, I was keenly interested in this so-called solar revolution, whatever it was supposed to be about. And as a red-blooded, patriotic American, I didn't like the idea of OPEC countries controlling our economy or our destiny. We had to find "alternative sources" of energy, whatever those might be.

Before I continue, let me say that if this was a book about energy policy, I could stop right here because very little has changed in the United States in the past twenty-five years regarding energy *policy*. Not even President Obama, who campaigned in part on a green-energy platform, could get an energy plan through his Democratic Congress in his first term. Today, we're debating "energy security" exactly as it was debated by us young hotshots over three decades ago in the bars and cloakrooms of the nation's capital, except that then we helped the administration come up with a national energy plan that was far more progressive than anything seen in the Congress since. Today, in the wake of the Gulf War, September 11, and the Iraq and Afghan wars, energy security is far more critical than it was during the Carter era. *Plus ça change, plus c'est la même chose.* Or is it *déjà vu* all over again?

And the environment? Few thought about it then. Today, we think about it a lot, but do nothing, while energy security has trumped concern over greenhouse gases, climate change, and global warming. The right of Americans to drive SUVs and gas-guzzling pickups trumped concern over the loss of young lives in Lebanon (258 marines were killed in 1981), Kuwait, Yemen, Somalia, Iraq (4,700 dead), and Afghanistan (more than 2,000 dead) as the United States seeks to keep the oil regions stable.

Meanwhile, back at the DOE: President Carter came down from Camp David, like Moses, clutching his national energy plan and proclaiming for all the world to hear, "This democracy which we love is going to make its stand on the battlefield of energy." He said that this would require the "most massive peacetime commitment of funds and resources in the nation's history." The $88 billion package Congress subsequently approved for Carter's energy initiatives was bigger than the Marshall Plan for the reconstruction of Europe after World War II.

On June 20, 1979, Carter gave his "solar energy" speech to Congress, outlining a "national solar strategy" that included the goal of supplying 20 percent of our nation's energy needs from the sun by the year 2000. Carter had announced his "domestic policy review" for solar energy the year before on "Sun Day" in Denver. Tom Tatum and I were ecstatic; everything we were working for had the full backing of the president of the United States, or so we thought.

Then, we discovered that more powerful people than our little group at the DOE had the ear of Energy Secretary Jim Schlesinger (who had been defense secretary under Presidents Nixon and Ford). These were, you guessed it, the oil companies. To Secretary Schlesinger's credit, he did say shortly after taking the job, and just after the near meltdown at the Three Mile Island nuclear facility, that "no nuclear plant will ever get built in the USA again," and until 2012 he was right about that. But that didn't mean he believed the president's goal of a 20 percent renewable-resource-based society was attainable, or that he'd throw the gigantic subsidies for nuclear and oil over to solar and renewables.

That $88 billion was largely going for the "alternative energy" called . . . *synfuels* (synthetic fuels). Does anyone remember synfuels? This was oil "mined" from shale rock by heating it in situ. Exxon and Occidental Petroleum raced to extract oil from the vast oil shale deposits of western Colorado. Exxon was going to strip-mine for oil, and Occidental drilled the world's deepest mine shaft, looking to bring it out that way. Did it matter to anyone that extracting oil from shale was more costly than any already-proven solar technology, thermal or electric? Not in a country ruled by "big oil."

"The old nuclear crowd and the fossil-fuel guys within the department were really astonished that the president made that kind of commitment to solar," Tom told me later. "The sandbagging started immediately. I tried to get the White House to allow the president's solar speech before Congress to be televised, but they said no, and it went by like a speeding bullet."

On the plus side, the Carter energy plan included substantial funds for Conservation and Solar at DOE, including research money for the new Solar Energy Research Institute (SERI) at Golden, Colorado, to be headed by the Earth Day cofounder Denis Hayes. We still had plenty of work to do on the solar front. We never believed in the "synfuel fix," and I wrote numerous articles exposing the program after visiting the huge "oil mining" operations in western Colorado. The oil companies spent billions of taxpayer money on synfuels, but President Ronald Reagan finally pulled the plug on it, calling it "corporate welfare"!

Synfuels were the precursor to tar sands, the enormously costly undertaking in Alberta that extracts oil sludge from rock with cata-strophic environmental consequences, and there is no politician on either side of the border with the courage to stop it and redirect our efforts toward sustainable energy alternatives, of which there are plenty, as you'll see.

Not all the oil companies got their hands dirty with oil shale: Atlantic Richfield (ARCO), under the leadership of Robert O. An-derson, sold its interests to Exxon. Anderson, founder of the Aspen Institute, saw synfuels for the scam that it was. His was the first oil company to invest in solar.

Since none of us was sure if President Carter would get a second term given his standing in the polls in 1979, we decided we had to act quickly if solar energy was going to see the light of day, so to speak. We had the blessing of the White House, even if the aging "old technology" guys in DOE were trying to kill us. I thought of German physicist Max Planck's quote, loosely translated as "New ideas in physics triumph only after the adherents of the old ideas have died."

The most exciting new idea in physics was photovoltaics, first devel-oped by Bell Labs in 1953; this new technology enabled the production of electricity from sunlight using "solar cells." Up to thirty-six of these small silicon wafers are soldered together with metal tabs and assembled into a "module," commonly called a solar "panel," which converts sunlight into electrons. The electrons flow as twelve-volt direct current from the solar cells. Asking me to explain this would be like asking a television network executive to explain how a TV works (see Appendix 1).

Paul Maycock, a physicist formerly with Bell Labs and Texas Instruments, headed up the PV shop at DOE, managing a budget of $900 million for applied research and "commercialization" under the Carter program. He was called Boomer, for he spoke in a voice that filled up a room of any size without amplification. Rotund, bearded, and intensely absentminded, he was busy bringing government support to the fledgling technology through contracts with manufacturers, research labs, and commercial innovators.

Paul told me, "Residential systems are being sold all over the world right now. Solar cells are already economically viable for off-the-grid, isolated homes and villages. By 1986, we expect photovoltaics to be fully economic for residential use in this country, at a cost to consumers of six cents per kilowatt hour in the sunnier parts of the country."

Paul was dreaming, and a little ahead of his time—by about twenty years—but his vision would eventually be realized.

Denis Hayes out at SERI in Colorado was predicting that we'd achieve DOE's goal of bringing the cost of solar cells down to seventy cents per watt by the mid-1980s. He too was dreaming. But by 2012 PV panel prices averaged less than a dollar per watt, a huge cost reduction from ninety dollars a watt, which is what solar modules cost in the mid-1970s. By 1980, the price for the PV panels used in the world's first 100-kilowatt array installed at Natural Bridges, Utah, was ten dollars a watt.

I was excited. Here was the technology that could save humanity! It was 100 percent American ingenuity. We could make enough solar cells from silicon, the most abundant element on earth after oxygen, to not only replace oil and coal, but to build decentralized, self-sufficient power networks managed by microprocessors that would allow every home to be its own utility. Suburban roofs would be covered with solar panels, making families energy independent; huge "arrays" of panels in the Southwest desert and installed on our commercial rooftops would produce industrial-strength power for the grid. Cars would all be electric, charged up by the sun (DOE's electric-car people predicted that nine million all-electric cars would be on the road by 1990). More dreaming.

Photovoltaics, along with wind energy, small and large hydroplants, biomass, cogeneration, energy conservation and efficiency, solar thermal plants, solar water heating, and whatever else the busy denizens of DOE's solar office were coming up with, would foster energy self-reliance and create a "decentralized" energy system.

America would be a land of "ecotopians," to borrow a word from Ernest Callenbach's visionary tract *Ecotopia*. A "solartopia" would follow.

Free energy from the sun sounded good to me. But we were way ahead of our time, and what we didn't know was that the technology to accomplish all this wasn't nearly as ready or as economically viable as we had hoped.

Tom Tatum asked me to help communicate this dream of a new future to the American people in what would become the largest solar energy communications campaign in American history. Actually, it was the only one.

Six foot three, charismatic, handsome, and irresistible to D.C.'s population of attractive single women, Tom focused on his calling like a southern preacher. He stood out like a Rhett Butler among the gray hordes of Washington bureaucrats, who seldom smiled and never laughed. He had fun, strolling by the sidewalk cafes of Capitol Hill with a different woman on his arm every evening, but by day his dazzling intellect was applied to only one thing: saving the world with solar energy.

Carter had given him carte blanche to try. Tom's one-of-a-kind Office of Institutional Liaison and Communications ("I made it up," Tom told me years afterward) meant he didn't need to work through DOE's Office of Public Affairs, which was controlled by ex-military types looking after the *real* concerns of DOE—processing uranium and making nuclear weapons. (The department was also known as "the bomb factory." When Carter formed DOE, Congress dumped the Atomic Energy Commission into it. The commission oversaw the government's breeder reactors, which manufactured fissile material for nuclear warheads, Atoms for Peace having somehow been forgotten along the way.) Tom was ready to take risks to launch a full-blown national promotion of solar energy, using all the budget at his disposal and damn the torpedoes.

We believed it was our job to communicate the new ideas, including how energy conservation could reduce the need for imported oil, what a solar economy might look like, and how we could attain it. There was an air of "national crisis" in Washington; the big worry was that oil could go to one hundred dollars a barrel. Such a price would destroy the world's economy. It had nearly hit forty dollars in the 1970s before retreating.

We had to raise the consciousness of the American people, and we had the platform and resources to do it from our base at DOE. What more did we need?

Hollywood! We'd use entertainment, television, celebrities, movie stars, and the cinema to get the word out, just like the government did to rally America against the Nazis in the early days of World War II.

But first we had to come up with a strategy. I suggested a high-level national retreat somewhere, and we found a glorious mountain camp above Boulder, Colorado, in the foothills of the Rockies. When the owner offered us the Gold Lake Ranch compound for free over beers at the Hotel Boulderado, we took it. A "Boulder hippie" with business acumen, he sought to operate a respectable mountain conference center, and he wanted to do his part for the solar revolution. The DOE would be his first big client. He only asked that we pay for the food. Tom OK'd the deal, and I was ordered to come up with an invitation list for the Gold Lake Solar Energy Media and Communications Strategy Conference.

We brought together young leaders and professionals from television, film, advertising, banking, politics, local government, architecture, publishing, construction, broadcasting, education, philanthropy, business and industry, as well as a former leader of the anti–Vietnam War movement. For the veterans of the antiwar movement, the No Nukes cause was the hot new crusade. However, our little DOE workshop was suspect among members of the No Nukes movement and the Hollywood rock music community, since it was the DOE that had promoted and subsidized nuclear power and also operated the country's two uranium-processing plants.

We met with the No Nukes people as well as with Musicians United for Clean Energy at their offices on Sunset Boulevard and explained, "No, we're not pronuclear. We're from the Office of Conservation and Solar, and we're reporting directly to the president." We reminded them that Carter's own appointee to head the TVA was against nuclear power. (David Freeman would continue that drumbeat two decades later, stating at a Sierra Club roundtable, "Nuclear power is dead, except in the hearts and minds of the religious believers in nuclear power. After September 11, we are surely not so dumb as to build more Trojan horses in our country. The danger of a penetration into a nuclear reactor—which is difficult but not impossible—is so horrendous that we've got to be out of our minds to build more nuclear power plants. And I say this as someone who's had as much experience with nuclear power as anyone in this country. In this age of terror, we just can't have them.")

Stopping nuclear power back then seemed like a fine cause, along with reducing the use of imported oil, but what was going to replace these energy sources? Well, the No Nukes people had been using John Hall's song "Power" ("Give me the warm power of the sun") to promote the alternative. A local musician sang it before the great fireplace at the Gold Lake Ranch following our welcome dinner, and we all dreamed of a new America powered by solar energy. (I wouldn't hear the song again until Peter, Paul, and Mary sang it from the stage at Earth Day 2000 in Washington, where actor Leonardo DiCaprio and Vice President Al Gore spoke passionately about the environment, standing beneath a two-hundred-foot-long banner proclaiming

"Clean Energy Now." Peter, a longtime friend, told me backstage just before the trio walked out to the mikes before the huge audience, "This is for you!")

The singer at Gold Lake was followed by a Ponca Indian poet named SaSuWeh, who recited his poem "Master of the Sun" and then offered around his peace pipe.

What were we smoking? I now wonder, looking back thirty-two years later. Actually, we were only drinking. No drugs were allowed at the retreat, so we had to be content with the clear Rocky Mountain air, French food, and large quantities of booze. I never knew who picked up the bar tab for what the subsequent DOE report called "an eclectic gathering of professionals."

The goals of the conference, as stated in the Gold Lake project report, were "to develop a communications strategy aimed at overcoming the institutional and psychological market barriers to solar energy in this country, and accelerate the transition to a renewable resources society including a 20 percent solar America by the year 2000; to form an advisory pool of non-governmental professional media resources; and to urge the formation of a private-sector national media coalition for the purpose of developing solar markets and prompting the use of the sun's energy."

What were we drinking?

Reading the DOE's sixteen-page summary report today is to sadly recall a dream unfulfilled. But Gold Lake *did* launch a national campaign to promote energy alternatives and energy conservation. And our Hollywood communications effort was a direct outcome.

I moved from D.C. to "duty station Hollywood," where I'd once paid my dues, like so many young writers, trying to get a feature film made (Tom would later move to Hollywood to make movies while I got into the solar business). Soon, Tom and I were meeting with all the big names in television series production, convincing them to help America solve its energy crisis. Gold Lake had opened a lot of doors. Energy-conservation messages were written into dozens of sit-coms, which reached tens of millions of people. I brought in a friend, producer Jon Davison (*Airplane, RoboCop*), who got director Ron Howard to help us, and Tom fronted a quarter million in

DOE money for *Reach for the Sun*, a TV film for children about solar energy and energy efficiency, coproduced with KCET-TV, Los Angeles. We worked with Robert Redford to help promote his "Solar Film," a seven-minute masterpiece of solar propaganda offering a new energy future. It ran in thousands of theaters that year as a public-interest trailer. Today, it is a curious artifact of a long-lost vision.

At Gold Lake we filmed a U.S. government public-service announcement featuring the Ponca Indian in a tepee teaching his small son about the benefits of the sun's energy. We intercut his solar paean with images of solar hot-water installations on Boulder apartment houses and wind generators in the Midwest. This was followed by dirge music over aerial shots of a supertanker making its way to U.S. shores with its cargo of liquid black gold. Then the words "explore solar energy, U.S. Department of Energy" appeared on the screen, followed by an 800 number for information. "Master of the Sun" is still viewable on YouTube.

Several thousand of these PSAs were distributed and played for years on hundreds of local television stations. We also produced a thousand radio PSAs featuring Hall's solar song. When the Reagan transition team swept through the DOE in the winter of 1980, they found this subversive material and burned the lot of it. President Reagan preferred the warm power of imported oil (he was right; it was cheaper to get it from Saudi Arabia than from oil shale or renewables). He also had no use for energy conservation. "America didn't conserve its way to greatness," he was fond of saying.

Actually, it was energy conservation and its counterpart, energy efficiency, that together accounted for an unparalleled energy success story arising from the Carter years. Five years after Jimmy Carter's one-term presidency, the country was using 15 percent less electricity than when Carter came into office, despite a healthy growth in the GDP along the way. This was attributable almost solely to government-sponsored energy-saving programs in partnership with industry, business, and consumers. The Office of Conservation and Solar instituted virtually all the then controversial programs that became commonplace in the decades after, such as using compact fluorescent lightbulbs, reducing energy waste in refrigerators and air conditioners, making buildings more efficient, implementing industrial cogeneration, and mandating household weatherization. The first solar tax credits proposed by DOE at this time were enacted by Congress, fostering a whole new solar water-heating industry. All this was Carter's unsung legacy.

Tom's shop at DOE led the way with public education, building national awareness for the need to "save energy." While we had not made any headway in "selling solar" to the American people, we were unexpectedly successful at selling energy conservation and efficiency. We brought in the mainstream press for special seminars inside DOE and worked with the White House to produce the first "energy fair" on the mall in Washington. We cajoled DOE's public affairs office to back a national campaign: "Energy: We Can't Afford to Waste It."

But on the solar front, we lost the war. After Gold Lake, Tom's office proposed to the White House a $50 million paid-advertising campaign for solar energy, but it was denied funding by the appropriations committees. By this time, Tom was getting pushed aside by the oil and synfuels people at DOE. They managed to get his independent communications responsibilities transferred to the DOE central public affairs office, which buried the solar message for good when a new energy secretary, Charles Duncan Jr., former CEO of Coca-Cola, brought in a military guy to oversee it.

"Duncan brought in nineteen special assistants, who were trampling all over everybody at the agency," said Tom, "and they spent a great deal of their time trampling over everyone at Conservation and Solar. All of them came from the Defense Department or industry, with no experience in energy at all."

Tom was incensed that a larger and larger portion of DOE's budget was going for the nuclear-warhead program instead of domestic energy needs. "This agency needs to be broken up," he said. Even *The Washington Post,* in an editorial, agreed. It never was. Reagan said he'd dismantle the DOE when he was elected. He never did. He couldn't. It remains a bloated bureaucracy that even enlightened secretaries like Hazel O'Leary, Bill Richardson, and Steven Chu could not tame.

At the same time, Congress continued to allocate sizable funds for solar and conservation. Paul Maycock's shop got $90 million in 1981 to support the "commercialization" (a term of bureaucratese I learned at DOE) of photovoltaics. The solar research center in Colorado got its operating money, but no money for public education. As a result, few Americans knew that the U.S. government was running a program devoted to solar energy that employed nearly one thousand people. The Solar Energy Research Institute was later renamed the National Renewable Energy Laboratory, excising the nasty word "solar."

Thanks to the Iran hostage crisis and the Middle East stew of unending violence, which was driving and controlling U.S. foreign policy even before the cold war ended, Carter was in trouble. Tom, by then getting further sidelined at DOE, saw the writing on the wall. He was disappointed that efforts to get the solar, conservation, and efficiency message out to the American people were being dismantled by DOE officials and that Carter's White House would not intervene.

Tom Tatum resigned from DOE in September 1980, citing "substantive disagreements" with the department's "balanced energy agenda." In his resignation letter, he pointed out his opposition to the enormous subsidies for synfuels technology. "I believe," he wrote, "the resources that have been arrayed to develop the synthetic fuels industry misallocate capital that is badly needed to implement comprehensive energy efficiency programs in the United States and accelerate the use of renewable resources to reach the President's goal of 20 percent solar by the year 2000."

Tom held a huge farewell party on the roof of the Hotel Washington in September 1980, attended by many of the city's young movers and shakers and dozens of the most beautiful and intelligent women in Washington.

Tom moved to LA. "There are even *more* beautiful women out there," he said. He married a lovely film editor, and Tom and I later collaborated on a movie treatment about a fictional conspiracy to stop a technical "breakthrough" in photovoltaics that would have badly hurt the oil companies. Needless to say, the movie didn't get made, although a feature film, *The Formula,* with a similar plot appeared two years later, starring Marlon Brando and George C. Scott. It bombed. Tom and I would not collaborate on solar energy for another thirty years.

Paul Maycock, head of the photovoltaics office, left DOE when the Reagan people took away his allocated PV budget. Had PV received the same push into the marketplace that nuclear power enjoyed, we'd probably all be living in sun-powered houses today. But Paul was told that, unlike nuclear and fossil fuels that were strongly supported with ten times the amount of subsidies that clean-energy technologies received at DOE, solar would have to make its own way in the market.

Our goal of a "20 percent solar-based society by the year 2000" had been a self-inflicted delusion. That goal during the following decades became an ever-receding horizon. In 2003, I sat through an amazingly boring roundtable workshop at the World Resources Institute in Washington, where representatives from numerous non-profit organizations and their earnest young policy wonks endlessly debated whether the "energy policy and environmental community" should call for America to derive "10 percent or 20 percent" of its energy from renewables by the year ... 2020! The group decided that it would only be realistic to unite behind a target of 10 percent renewables by 2020.

That year, the press reported that "despite pleas from a bipartisan group of 53 senators, the Republican House leadership eliminated a provision from Pres. Bush's energy bill that would have required our nation to generate 10 percent of its electricity from renewable resources by 2020." At least Carter had gotten his 20 percent goal included in his energy bill twenty-five years earlier, largely based on the "energy security" argument. Presidential candidate John Kerry said in a speech in January 2004, "I support a national goal of producing *20 percent* of our electricity from renewable sources by 2020." *Plus ça change ...*

Such goals are nearly meaningless. The good news is that we may actually exceed these once ambitious targets given the rapid trajectory of solar and wind energy in America. That is the paramount message of this book.

Paul, having left government service, launched *PV News* in his basement. Exactly ten years later he joined the board of directors of my nonprofit organization and later my first commercial solar venture. Twenty-five years after Tom and I left the DOE's solar energy office, our attempt at fomenting a solar revolution having failed, a book reflecting the realities of solar power in the new millennium appeared, written by Travis Bradford, called *Solar Revolution,* but *without* the question mark as in the title of this chapter. Travis also purchased Paul Maycock's *PV News,* transforming it into the PV industry's leading insider newsletter.

I went back to Colorado to escape the Reagan years, continued freelance writing, and became the marketing director for a large ski resort. Later, I became a real estate developer of sorts, buying and rehabilitating a historic building and running several businesses. I didn't think about solar energy for nearly a decade.

Before working for the DOE, I had written the cover story for the premier issue of *Outside* magazine in 1977 at the invitation of Jann Wenner, editor of *Rolling Stone* magazine, who launched the new outdoor sports publication. The first issue also contained a feature about the "eco-commando" environmental organization Greenpeace, which I had not heard of, founded in 1971 by a small antinuke group in Vancouver. It stuck in my mind.

I missed not having a mission. Not willing to grow old in a ski town, I landed back in Washington in 1987 to take a low-paying job at Greenpeace USA as their national director of media and public relations. Here I met Greenpeace's chairman, David McTaggart, a highly energetic Canadian who had lived and worked and developed real estate in California and Colorado ski resorts before heading off to the South Pacific in a small boat, where he managed single-handedly to end French nuclear testing in the atmosphere. This was a career switch if there ever was one! We had a lot in common. I liked David immensely, and years later I visited his magical house in an ancient olive grove in Umbria, where he asked me how to put solar hot-water and solarelectric systems on his roof. Before we could proceed, he was killed in a car crash in Italy, a catastrophic loss to the global environmental movement.

It was at Greenpeace that I first heard the phrase "global warming." I learned a great deal at this amazing organization, which was "campaigning" on just about every environmental issue there was, from nuclear testing (belowground) and nuclear power to toxic wastes, cleaning up rivers and oceans, saving marine mammals, ending the seal slaughter in Canada, and stopping the hunting of whales.

In my new job I was determined to do everything possible to make Greenpeace as famous as it deserved to be. When the French government sank the Greenpeace antinuclear campaign ship *Rainbow Warrior* in New Zealand in 1985, the organization's profile soared in Europe. I wanted to raise its visibility in the United States.

I called Tom Tatum in LA, where he was producing a feature film about motocross racing and brought him back to D.C. so we could coproduce *Greenpeace's Greatest Hits,* a video featuring the organization's first decade of activism. I asked the late John Forsythe, the star of *Dynasty,* who possessed perhaps the greatest voice of the twentieth century, to narrate it. This was all great fun, and the video was a success. However, I was concerned that the young activists at Greenpeace seemed to see the world in stark, even hopeless, terms. Greenpeace was about highlighting problems, not providing solutions or hope. And solar energy as a positive partial solution to the energy and greenhouse-gas-emission problem, and hence to global warming, was not on its radar screen at the time, although that would change.

Five years later Greenpeace International, based in the Netherlands, launched a solar campaign. Geophysicist Dr. Jeremy Leggett, of Oxford University, advised the group on the need for radical change in energy policy. Our paths were to cross in 1997, when we both launched solar companies.

Almost nine years had gone by since I left DOE's employ. And I had begun to wonder, Whatever happened to solar energy?

I was looking for hope, so I left Greenpeace and began consulting for America's largest solar manufacturer, Solarex, based in Frederick, Maryland. It was headquartered in a large slab of a building, clearly visible from the interstate, and was dramatically covered on one side by three hundred kilowatts of blue polycrystalline PV modules gleaming in the sun. Started in 1983 by Drs. Peter Varadi and Joseph Lindmeyer, who shared a vision of "bringing solar down to earth," Solarex was producing solar modules for terrestrial use as well as for our space program, which is where PV got its start. The terrestrial market, then, was growing by 30 percent a year.

During the Reagan years, the oil companies had invested heavily in solar PV, and the industry was progressing quite nicely. Solarex was later sold to Amoco, and it eventually became BP Solar when British Petroleum purchased both Amoco and Standard Oil of Ohio. We will hear more about BP Solar later, before its demise in 2011.

I had no idea at the time that my solar odyssey had only just begun, as you'll see in the following chapter.

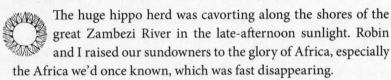

2

A WORLD OF DARKNESS: 1990

The huge hippo herd was cavorting along the shores of the great Zambezi River in the late-afternoon sunlight. Robin and I raised our sundowners to the glory of Africa, especially the Africa we'd once known, which was fast disappearing.

Robin Moser, a third-generation South African of Swiss-Irish descent, said, "We used to drive here from Jo'burg to Vic Falls to take the BOAC flying boat to London. There was no other air service, and this was the only river big enough for the four-engine amphibious plane to land. From here it would fly to Lake Tanganyika, then to Egypt and the Nile, then on to Europe."

I had my own memories of the old Africa. I had motorcycled from Cairo to Cape Town in 1962–63 and had fallen in love with the continent before most of it fell to slaughter, famine, and corruption. I had met Robin quite by chance as we stood on the high steel-arch bridge planned by Cecil John Rhodes and inaugurated by the son of Charles Darwin in 1905, which stretched across the Zambezi below Victoria Falls (and is today a favorite bungee-jumping site).

"You've come on *that* all the way from Cairo?" the young man had inquired, looking at my 250 cc, two-stroke motocross bike. Robin was visiting the falls with his family. We became friends, and he invited me to stay at his home when I later got to Johannesburg. We remained friends over the years, and now, nearly thirty years later, I had come back to Africa for the first time and had looked him up. He immediately suggested we revisit Vic Falls together for a couple of days. We booked an Air Zimbabwe flight to Bulawayo, rented a car, and took rooms at that grand old Victorian pile, the Victoria Falls Hotel. Robin made sure I brought a jacket and tie for the dining room. "They keep the old traditions there," he said.

But this was not just revisiting the past, I was looking to the future and what it could bring.

I believed South Africa had a vibrant future. I always knew it would settle the apartheid problem, and I had the feeling that 1990 would be *the* watershed year for profound changes. I wanted to be there when history was made.

Meanwhile, Zimbabwe, formerly Southern Rhodesia, had emerged as the "hope of southern Africa." Robin wasn't convinced about this new, emerging Africa run by the black majority, and this was clear when he saw the two African pilots enter the 737 cockpit at Jan Smuts Airport. He almost got off. On this particular flight the Boeing happened to make the hardest landing I had ever experienced, and I thought the wings would come off. "See," said Robin, "these people aren't ready."

I disagreed. "Ready or not, it's their country, and pretty soon they're going to be running *your* country, too."

I was looking at Africa now through very different eyes, at a very different time. The "winds of change," to borrow a phrase from Harold Macmillan's controversial 1960 speech before the South African parliament, had at long last blown new hope across the whole of southern Africa.

I wanted somehow to make a small contribution to this new "developing" third world, or what Paul Hawken, the author of *The Ecology of Commerce,* calls the two-thirds world. In the poorer countries, where the majority of the world's people live their short, happy, but often desperate lives, lay the great challenges for the planet. How would they develop? How would they acquire even a small measure of the lifestyle we take for granted? How could they have even such a basic thing as electricity without destroying the global environment?

Driving through the bleak Hwange coal country, Robin told me Zimbabwe and South Africa had more coal deposits than the entire world could ever burn. "Great," I thought, watching the huge Hwange power plants spewing clouds of gray smoke into the African skies, trying to envision what the air would be like when black people, as well as white people, were able to get electricity. Only 2 percent of rural Zimbabwe, where 90 percent of the people lived, had electricity. Ninety-eight percent of Zimbabwe's electricity was used for mining and industry, not for lighting homes, schools, or businesses. Kerosene and candles lit most of the homes in Africa—and most of the homes in the two-thirds world. According to United Nations statistics at the time, two billion of the five billion people on the planet had no electricity. Today, there are seven billion people and even more families without electricity.

All this coal! All these people without power! Everyone looking to a brighter future, figuratively and literally. And hydropower generated from the mighty Zambezi at Kariba, the enormous high dam I'd watched Italian workers constructing in 1962, was not the answer because the electricity had to be *distributed,* which cost more than generating it. Besides, the massive hydroelectric projects of the 1960s had become their own environmental nightmares, displacing people, destroying ecosystems, and even altering microclimates. Dams and coal were definitely *not* the future.

At the Victoria Falls Hotel that night, I couldn't sleep. It wasn't the screech of the night insects that I could hear through the window, nor was it the ceiling fan's breeze on my face. I was having a Big Thought.

Why not use photovoltaics to bring basic electricity to these people? I wondered, lying there in the Mzilikazi Suite, listening to the distant roar of the "smoke that thunders" outside my window. It was as if a light had gone on in that room, although it was as dark as only the African night can be.

Solar energy, which seemed at the time to have little chance in America, as we'd learned the hard way at the Department of Energy, could be the best new energy source for the widely dispersed population in the developing world. And the environmental threat posed by the increased greenhouse-gas emissions that would be released as the developing world developed could be partially avoided.

As a consultant to Solarex, then America's largest solar PV manufacturer, I'd learned that its biggest market for photovoltaic modules was South Africa, a fact they didn't advertise during the period of the antiapartheid boycotts. PV was widely used for telecommunications, wireless telephony, radio and television broadcast translators, railroad signaling, lighting for remote post offices, and power for police stations. South Africa was the perfect crucible for this new technology because it was, as the tourist brochures said, "a world in one country." It was the "first world" and the "third world" in one place, a world of vast distances where whites all had electrical power, most blacks in rural areas didn't (blacks in the townships did), and white entrepreneurs and businesspeople saw a cost-effective "application" for solar electricity.

The government was already putting solar energy to use in South Africa in a big way, and a few small Afrikaner-owned businesses were actually selling solar home-lighting systems to black farmers, who could pay for them on "lay-by" plans if they didn't have the upfront cash.

Before returning to the United States, I flew to Cape Town. I had a hunch that something momentous was about to happen.

Prime Minister F. W. de Klerk had hinted that sometime that year Nelson Mandela would likely be released from prison, where he had proudly kept the dream of the African National Congress (ANC) alive since 1964. My white South African friends could not and would not believe it. Robin had driven me in his Rolls-Royce to his weekend farm in the Magaliesburg. The transmission line to his country house bypassed thousands of black homesteads along the way. I tried to talk to him about Mandela, but he wouldn't have it. "No way they're going to be allowed to run this country," he said, echoing British prime minister Margaret Thatcher, who had said in 1987, "Anyone who thinks that the ANC is going to run the government of South Africa is living in Cloud Cuckooland."

Thieves kept breaking into Robin's weekend retreat to steal his stereo. "I just buy another one," he said, nonchalantly.

"How do they power them in their huts without electricity?" I asked.

"They hook them up to batteries. They steal those, too."

"Why don't you just buy your African neighbors stereos and arrange for electric lines to their houses," I inquired. "You could afford it." He gave me a dour look. I suggested that solar photovoltaics might help.

Now, in Cape Town, I was watching a country in transition. I climbed atop the city market building at City Hall Square, which lay before the grand edifice designed by the prolific colonial architect Herbert Baker. Violence broke out. Sixty-nine people were shot, four killed, on that long hot February day under the searing cape sun. I shared my binoculars with polite young black South Africans, who were having as hard a time as my friend Robin believing that what was about to happen would actually happen. But it *was* happening, right there before our eyes.

I could see them plainly through my binoculars, and suddenly everyone around me needed to have a look. Nelson Mandela, with his wife, Winnie, appeared on the steps of City Hall before the mixed-race crowd of eighty thousand expectant supporters who'd waited, as I had, almost six hours in the blazing southern sun. Mandela was free, and so was South Africa, and so was the human spirit that believed a better world could be made here and elsewhere! Maybe it wouldn't be, maybe it can't be, but it never hurts to believe, and symbols of hope like Nelson Mandela are God's gift to mankind. The "winds of change" had finally blown clean through this beloved country. With the highest solar radiation in the world after Saudi Arabia and Australia, South Africa was now a country where I could legitimately pursue my own dream of helping to bring solar power and light to rural people still living in the dark.

Back in Chevy Chase, Maryland, I took over our dining room table to form an organization that could promote solar power around the world and help bring it to those who needed and could use it. No solar company had yet made a profit, so that approach didn't bode well. I thought about setting up a nonprofit organization, and that got the following response from my friends at Solarex: "Sure! We're a nonprofit, too. Your not-for-profit organization and our profitless company can work together to sell solar to all those folks in these countries who don't have any money. Great idea!"

Because I didn't know what I was getting into, because no one else was doing this (besides one Richard Hansen in the Dominican Republic, whom I'll get to shortly), because I simply had nothing more important to do, and because friends thought it was a great idea, I decided to set up a nonprofit organization to promote and finance solar rural electrification in the developing world. My old friend Oliver Davidson at the State Department's Office of Foreign Disaster Assistance thought it was "a good idea," and he had the international development credentials to back up his opinion. I had also consulted Dr. Peter G. Bourne, assistant UN secretary-general and former special advisor to President Carter, who strongly supported the idea of an NGO (nongovernmental organization) devoted exclusively to bringing solar power and light to the developing world.

I thought long and hard about a name, and in the early summer of 1990 I came up with *Solar Electric Light Fund*. I borrowed the name from Thomas Edison's Edison Electric Light Company, which pointedly revealed that the original electric power companies were set up to deliver one thing only: *electric light* for houses, business, and industry. They were all "light" or "electric lighting" or "illuminating" companies. I would just add the word "solar." The acronym was SELF, for energy SELF-sufficiency and SELF-reliance.

The Solar Electric Light Fund was officially launched on June 23, 1990. Its mission was to bring solarelectric lighting to rural people in the developing world who had no access to electricity. We called it an "electrifying idea."

Patricia Forkan (my wife and then the executive vice president of the Humane Society of the United States), Ollie Davidson, and I formed the founding board of directors. I'd known Ollie since Ohio, and we had been in Vietnam together—I as a freelance correspondent, and Ollie as a civilian "district advisor" for the U.S. Agency for International Development (USAID) in Hau Nghia Province, near the Cambodian border. Ollie had first introduced me to exotic lands through his talks and slide shows in Ohio, where he presented his backpack travel adventures of his college years. Long before the world was overrun with "backpackers," there was Ollie, journeying through the Amazon or up the Nile with his army-surplus rucksack. Before USAID hired him to join America's misadventure in Vietnam, he would return to remote villages in West Africa and along the Amazon, where he'd earlier made friends during his backpacking days, to bring them medical supplies. No one I had ever met cared more for the common people in the poorer countries than Ollie, and I was honored that he not only thought the Solar Electric Light Fund was "a good idea" but that he agreed to be a director. Peter Bourne joined the advisory board and opened doors for us at the UN and United Nations Development Program (UNDP).

Paul Maycock, retired from the DOE, also agreed to join the board. Having run the DOE's PV program for Carter, he was now regarded as the most knowledgeable expert on the worldwide PV industry. Paul told me, "As I get older, I find myself more motivated by the role photovoltaics can play in providing subsistence levels of electricity for the two billion people of the world who have no electricity. These people have little chance of ever seeing electricity in their homes, clinics, or schools." He understood the problem, and the solution.

Now I had to raise money. Solarex's CEO made the first grant of five thousand dollars, even though he skeptically wondered how the hell I would ever figure out how to sell solar electricity to poor people in the developing world. "It's a huge market, but these people don't have any money," he reminded me one more time.

I didn't have any money either, and didn't even have a salary or an office. With the Solarex grant I bought my first computer, a Zenith laptop with an 800-kilobyte memory, and began writing grant proposals. To my great surprise, a couple of small foundations specializing in "environmental issues" came through with support, and SELF was off and running. I opened a one-room office in a third-floor walk-up on Connecticut Avenue, near Dupont Circle, and installed a cheap air conditioner. Tom Tatum from DOE days hosted a small fund-raiser in the Hollywood Hills, film producers and TV actors dropped by, and we raised one thousand dollars. SELF maintained this link for the next six years, and some of Hollywood's most famous and committed environmentalists became regular supporters. Activist actor Ed Begley Jr. joined SELF's advisory board and hosted a fund-raiser at his house, where I met the original *Tonight Show* host, Steve Allen, whom I'd grown up watching in the fifties, when he was America's biggest TV celebrity. He was keenly interested in our plans to bring electricity to Chinese peasants, since he had traveled a great deal in China and had written a book on the country with his wife, actress Jayne Meadows.

Speaking of Hollywood, we were inspired by actor Jack Nicholson's interview for an environmental magazine in which he said the only hope for mankind was solar electricity. "Solar electricity, solar electricity, solar electricity," he was quoted as saying. "There, I've said it again." Many years later, Larry Hagman, J. R. Ewing in the TV series *Dallas,* joined SELF's board of directors, and remained a big supporter of SELF until his death in 2012 (he had over 80 kilowatts of solar on his California mountaintop home).

Back in D.C., I bought a fax machine, which was a relatively new gadget in 1990. Before e-mail, fax machines shrank the world. This was the miracle device that allowed inexpensive communication beyond the normal borders of international telecommunication, and when messages began to arrive from Baluchistan (yes, Baluchistan—a province in southern Pakistan) and other exotic places, I felt like I'd been admitted to media theorist Marshall McLuhan's "global village."

McLuhan, the visionary Toronto professor who prepared us for a world made smaller by electronic communications, had inspired me more than I realized. I found my original paperback of his book *Understanding Media: The Extensions of Man,* published in 1964, in which I'd underlined that same year—the year I met him at the University of Colorado following a speech—"Light is a non-specialist kind of energy or power that is identical with information and knowledge. . . . Grasp of this fact is indispensable to the understanding of the electronic age. . . . *The electronic age is literally one of illumination*" (my italics). McLuhan had always been my prophet, but I didn't really know why until I caught up with him thirty years later, after he died, and reread this brilliant thought: "The electric light ended the regime of night and day, of indoors and out-of-doors. . . . In a word, the message of the electric light is total change. It is pure information without any content to restrict its transforming and informing power."

More grants came in, and I began looking for countries where I could apply my concept of small-is-beautiful solar power for householders. Serendipitously, the countries found me. They usually reached me by fax, or responded by fax if I contacted them first by mail. The first was, in fact, the island of Serendib, once known as Ceylon, which changed its name to Sri Lanka in the 1970s. The word "serendipity" was coined when an eighteenth-century English writer found himself in this equatorial island paradise. It means, according to modern dictionaries, "the faculty of making happy and unexpected discoveries by accident."

This was to be the story of SELF over the next seven years. Our quest to bring solar power and light to remote villages in eleven countries was serendipity squared. But before heading off to Sri Lanka, I needed to do a little more research.

While investigating the possibilities for this unique venture, I had arranged to meet with a young World Bank researcher who had published a study on large-scale and minigrid solar power installations in Pakistan. These were all multimillion-dollar projects featuring "central power arrays" of PV modules, which were connected to individual homes. This was my first encounter with the World Bank, about which you will hear more later. The wholesale failure of these projects, described in the report, had given PV a bad name as a development tool.

More than anything, reading this study convinced me SELF was on the right track. Most engineers had not been able to conceive of using PV modules in a small and decentralized way, house by house. Thus, the companies like Solarex; ARCO Solar, in California; and Total Energie, in France, a division of the oil company, had foisted their highly engineered, centralized PV megaprojects on the various international development agencies, including the UNDP, USAID, and individual country donors. All were disasters.

They failed, according to the bank's report, because people stole the copper wires that connected the houses to the PV arrays. And then people stole the PV modules because they wanted to put them on their own houses to charge batteries. They failed because the power in the batteries could not be "managed" and delivered equally to all the households, despite sophisticated electronics, which tended to break down in any case. Before any other power solutions were offered them, the local people were already using car batteries, charged once a week in town, to run their TVs. They also knew that *one* 40-Wp PV module could supply their house.

Thus, a two-hundred-module array of 100-Wp modules, engineered to supply Western levels of electric service to twenty houses and to satisfy the design parameters of Western engineers, could in fact serve two hundred houses, one by one, with the subsistence level of power the houses actually needed. The technical specialists who dominated the solar industry at the time could not grasp this, and the procurement managers, project developers, and local commission agents were not interested in anything but large-scale, Western-style projects. The solar companies had no way of selling their product to individual households; they wanted megadeals, paid for up front by the United Nations and by various developing-world governments.

A point of clarification here: A solar "cell" is three to five inches wide. Placed together, six to thirty-six solar cells become a "module." Multiple modules form a "solar panel," and multiple panels become a "solar array." The largest solar modules at the time produced 80 to 100 watts of power in peak sunlight, hence "Wp." Today, "solar panel" is the popular name for a solar module, a term only used by the industry. Today, 280- and 300-Wp panels are commonplace (see chapter 4).

There were two technical considerations that had to be addressed. First, solar PV produces only direct current (DC), and DC power, as Thomas Edison discovered, doesn't travel very far. Thus, centralized PV systems are inefficient because they must send their power hundreds and even thousands of feet over very thick wires, resulting in

SELCO-India technician shows PV panel to Indian villagers. *(Philip Jones Griffiths)*

large "line losses." In the 1990s, efficient "inverter" technology, which converts DC to AC power, was in its infancy. Second, solar power can only work in the daytime, so power must be stored in batteries at night, and how do you control how much each home uses each night? Many engineers have tried to solve this problem, but have always failed.

The bank's report did not dismiss PV as a development tool; instead, it strongly advocated stand-alone solar systems. "It is shown that a decentralized approach for household PV systems in which individual households or buildings are powered by individual PV systems is less costly than a centralized approach in which a village is serviced by a single PV array and mini-distribution system," the report stated clearly. It concluded, "PV is clearly the least-cost technology for off-grid village/household power supply." That's what I had thought, based on my own initial research, and that was exactly what we had set up SELF to do. I was delighted to see the experts at the World Bank concur, basing their opinion on expensive, detailed analysis and evaluation.

* * *

But I still needed "proof of concept," as they say, since I had not yet seen an actual house powered by its own small solar panel. I heard about one engineer who understood the basic simplicity of stand-alone DC power produced by individual solar PV modules, installed house by house, each with its own battery, avoiding transmission costs and power losses. He was Richard Hansen, who had been working in the Dominican Republic for the past four years, install-ing "solar home systems" (SHSs) through his Massachusetts-based nonprofit group, Enersol.

A former Westinghouse engineer, Richard had fallen in love with the "DR" during a vacation there and returned one day to marry a Dominican. As an engineer who had studied renewable energy sys-tems, he saw the potential for solar PV as a useful technology to make people's lives better. Since the DR had a pitiful electric utility that could not deliver power beyond the main towns and that suf-fered frequent blackouts because the government could not afford to buy the imported oil needed to run the generators, Richard decided to do something about it.

He brought in several solar modules, hooked them up to batter-ies in a house he rented, and invited the locals to see it. The ability to have bright electric lights at night, watch a small TV, and listen to Dominican salsa on their radios by day created an instant demand for solar systems. Some families spent ten dollars a month on radio batteries in order to listen to music, which to Dominicans is almost as important as food. They spent even more money on car batteries, which they hooked up to small black-and-white TV sets, and an ad-ditional sum on kerosene for lighting. For the same money they could buy an SHS, or they could buy it on monthly installment payments if credit was available.

Richard formed a U.S. nonprofit, found a few private donors, got a two-thousand-dollar grant from USAID, and set up an office out-side Puerto Plata on the north side of the island. The Martinez fam-ily bought the first system on sight with a down payment and a three-year loan from the revolving fund Richard set up. Solar finance was born. By 1990, Enersol's community solar program had financed over one thousand customers. Over the years, Richard hired half a dozen ex–Peace Corps volunteers to help in his quest.

I contacted Richard and was offered a chance to visit Puerto Plata to meet him and see his project. I arrived at the two-room airport in October 1990, and Richard collected me in his jeep. We immediately went searching for gasoline. A lack of foreign currency prevented the government from paying tanker captains for their oil shipments, so the oil ships sat offshore, waiting for dollars. Naturally, there was no electricity either, since their power plants ran on diesel fuel.

The DR, with seven million people, is a microcosm of the problems of the two-thirds world: high birthrates, no resources, terrible corruption, worthless currency, horrible climatic conditions—and hurricanes every few years that set the island back decades. I'd traveled the world and been all through Mexico and Central America, but I had never seen poverty at the level it exists in the DR, a country one hour's flying time from Miami, with more people than Switzerland or Greece (not counting the million Dominicans who—no wonder—live in New York City).

Nonetheless, the rural areas were gorgeous and unspoiled, and they offered promise to hardworking farmers, who earned a decent income and built sturdy houses, which they painted in bright colors. These were Enersol's customers.

"We let the sun distribute the power, not copper wires," Richard pointed out. He took me to visit two dozen houses up a remote valley reaching into the tropical hills. Every house had music playing, many residents had bought black-and-white televisions, and everyone was happy to have electric lights at night, including the owners of the small shops along the way. Nothing is darker than a rural village without electricity, and one 12-watt DC fluorescent bulb (equal to a 60-watt AC bulb) looks like a prison-yard spotlight in that penetrating blackness.

Some of the wealthier families had bought DC juice blenders and even large stereo sets. Everywhere, children were smiling, and the families invited us in for fruit or a soft drink. At the end of the day, Richard took me back to his own house, which he'd built himself, a two-story concrete- and cement-block affair with wraparound

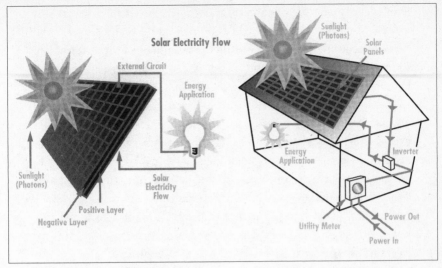

How a grid-tied residential solar power system works *(Spire Corp.)*. A stand-alone system uses a battery without a grid connection and a charge controller instead of an inverter.

veranda and solar-powered everything: hot water, computers, TV, radios, stereo, water pumped from his own well, and lights everywhere. He had a kerosene refrigerator because DC fridges were still expensive.

We sat in his office and studied revolving loan funds. This was the key. Without credit of some sort, which we in America took for granted, people could not afford these solarelectric systems. Richard had pioneered the idea of revolving credit funds, but he had not had much success finding the funds to revolve. He wasn't a bank.

Meeting Richard Hansen further convinced me that SELF was on the right track. Back in Washington, as a way to say thanks for the inspiration he provided me, I found some additional financial support for him from USAID. Then, I set about raising money for SELF so we could embark on our first project, which, serendipitously, had emerged from Sri Lanka.

I had been to Sri Lanka in 1972, when it was called Ceylon and when it was still a true paradise, before the ethnic and political civil wars started, before the devastating deforestation that altered the monsoon patterns and caused droughts and killer floods, and well before the 2004 tsunami that killed 35,000 people. Sri Lanka, 270 miles long and 140 miles wide, is one of the most beautiful places on earth. It has thick jungles and open plains, meandering rivers and high mountains, ancient cities and spectacular ruins, wild elephants and tame ones, and lots of other wildlife. Its 21 million inhabitants are the best-educated in the developing world. The great majority are Sinhalese, mostly Theravada Buddhist, while 15 percent are Hindu Tamils, and there are small populations of Muslims and Christians. Ten percent of the population speaks English.

Sri Lanka was among the first of the British colonies to achieve independence, in 1948, and it was the first former colony to give women the vote or to have a woman as prime minister.

In Moratuwa, thirty kilometers south of Colombo, I met our project partner-to-be, Dr. A. T. Ariyaratne, president of the Sarvodaya Shramadana Movement, to whom I had been introduced by the W. Alton Jones Foundation in Charlottesville, Virginia, which offered funds to SELF to set up a pilot solar program. (W. Alton Jones was an oilman whose heirs were trying to do good things with the family money by supporting unique environmental initiatives, including our plan to bring electricity to poor people without producing more greenhouse gases.) One of the largest NGOs in South Asia, Sarvodaya was founded in the 1950s by Dr. Ariyaratne, or Ari as everyone called him. Sarvodaya means "universal awakening." Built from the grass roots up, it focused from the beginning on what Ari called "people-centered development." From a village-based "shramadana" movement (meaning "sharing labor and energy"), Sarvodaya had grown to a membership of over two million people, with development activities in eight thousand villages. To many people, Dr. Ari was the contemporary Gandhi of South Asia, equally admired in India and other countries in the region.

Dr. Ari welcomed me with a great hug, despite his diminutive stature. Then, he lit up the first of many cigarettes. We entered his inner chambers at Sarvodaya headquarters, a sprawling compound

Sarvodaya's Dr. A. T. Ariyaratne ceremoniously flips on electric light at first solar-powered schoolhouse in Sri Lanka, installed by SELF in 1992. (*Author*)

at Moratuwa. On the wall were photos of Dr. Ari with numerous heads of state and at global religious and development conferences. I was immediately struck, not by how famous he was, but how humble: a more down-to-earth, easygoing person I couldn't have imagined. But his energy was palpable, as was his charisma.

"So we're going to do a solar project together!" he said. "I've been a long believer in solar energy and new technologies." He presented me with copies of a half dozen books he had published about rural development, in which he offered his practical views on how to lift people from poverty and illiteracy. He believed in practice, not theory. Dr. Ari confounded his international donors by having as his only "theory" the belief that poor villagers, if given the chance and the tools, could lift themselves out of poverty.

One of his tools was self-financing, and he had established

Sarvodaya Economic Enterprises Development Services (SEEDS), an island-wide credit union that made small loans to hundreds of thousands of families who had never been inside a bank. This was long before microcredit became fashionable. We were going to set up a "solar fund" within SEEDS that could self-finance the purchase of SHSs (solar home systems).

First, Dr. Ari drove me in his Mitsubishi Pajero on a dark journey through the jungles to visit remote communities eking out a subsistence living growing coconuts, cashews, and rice. It was nighttime, and there were no lights anywhere. Small bottle lamps illuminated the rough daub-and-wattle houses with their flammable thatch roofs. People were thrilled to have a foreign visitor and proud to show me their simple lamps and lanterns.

Arthur C. Clarke, author of *2001: A Space Odyssey,* who had lived in Colombo for thirty years, later told me that ten thousand people a year were maimed and disfigured by kerosene-lamp fires in Sri Lanka. This was one reason he was such a big supporter of sun power. He had written, "The village in the jungle can leap-frog into the modern age, thanks to solar electric conversion devices."

I was confronting directly the poverty of the two-thirds world. "Can we help these people with solar power?" Dr. Ari asked me. I tried to figure out how much money these simple villagers had, or earned, but that was not an easy question to ask and, if asked, was not likely to get an honest answer. However, it was clear that without some kind of financing assistance, and without low-cost but reliable solar home systems, these people had no hope of moving up to electric lights. They didn't expect the government was going to bring them electricity anytime soon. Without solar lighting, they would remain in the dark.

When I visited the offices of the Ceylon Electricity Board, I learned that 70 percent of Sri Lankans remained without power and that the barriers to their getting a grid hookup were money and generating capacity, which was limited by how much rainfall filled up the hydrodams. The electricity-board officials told me a coal-fired generating plant was planned, the island's first, which would have to *import* coal from Indonesia. Sri Lanka had only hydroelectric plants that used a so-called renewable resource that was not always sure to be renewed.

There, at the front lines of rural development, stood Dr. Ari, seeking solutions and looking at solar photovoltaics for the first time. Over the next five years, we did build a successful solar program in Sri Lanka, which became the base for the largest solar project financed by the World Bank. What I learned in Sri Lanka, after endless meetings with World Bank economists, power-company officials and planners, village-based Buddhist leaders, and local environmentalists, was that we could deliver an SHS for less than it cost to run an electric line one hundred meters from the nearest power main. And the increased cost of the electric line included only the hookup and house wiring, not the cost of transmitting or generating the power. The sun is, indeed, a better distributor of power than copper wires, especially when houses are widely dispersed, as they are in rural Sri Lanka and in most developing countries. Sri Lanka became our solar learning lab.

The Solar Electric Light Fund was now off and running. I still had no staff, but grant funding was coming in for projects in China, South Africa, Nepal, Vietnam, and an additional project in Sri Lanka. For between $30,000 and $50,000 in project grants, it was possible to launch a pilot project of a few hundred households in these countries. "Core support" grants covered travel expenses, office rent, and my modest salary. This allowed SELF to use project grant money in its entirety for each project, without deducting administrative costs. I intended to give the environmental grant makers who had taken a chance on SELF the biggest bang for their buck.

SELF was like a guerrilla nonprofit—fast on its feet, hard to see (since all the projects were at least ten thousand miles away), cost-effective, and decisive. We got results because of all the exceptionally honest, hardworking, and dedicated people I was able to find in my travels to a dozen countries.

Maybe it was because I was older or just a simpler person and had not grown up in Washington's heady, self-important, deadly serious environment, but I had little patience with the way some things were done in the capital's nonprofit community—especially with regard to international "programs." I wanted simple, practical, working *solutions*. I also wanted to see hands-on, tangible results; something that directly affected people; something I could photograph! It didn't seem we were really making a difference unless human beings benefited directly. SELF soon acquired a reputation for taking action and getting results with very little money. I learned that I could establish, through actual example, a national solar-electrification program in a country with less than the cost of one World Bank consulting mission, or less than a quarter of the salary and expenses of one USAID officer.

I was soon having a wonderful time, meeting so many terrifically interesting and entertaining people around the world as I traveled to their countries. I was invited by the UN Development Program to head a UNDP–World Bank mission to Zimbabwe to design a national solar program. I was invited to South Africa to come and electrify a village in the Valley of a Thousand Hills. I was invited to China by a woman I met at a solar energy conference in Denver in 1992. A group of Nepalese professionals invited me to Kathmandu to help them launch a rural solar program. And the Vietnam Women's Union, an eleven-million-member organization, invited me to help them bring solar power and light to their poorest and remotest constituents.

These projects—covered in the following chapters—all meant years of hard work, but they all got "implemented" successfully and had "impact," in the jargon popular among development professionals. Of the twelve countries we worked in, it was only in Zimbabwe that SELF was not able to actually implement the project, because we had designed it. (Under UN rules, the project developer cannot implement or manage it.)

I learned a great deal putting the Zimbabwe project together and was pleased when it was passed and budgeted for $10 million at a meeting at UNDP headquarters in New York, where I had to defend my "planning document" before a dozen UN and UNDP officials. Under the direction of Gibson Mandishona, whom I hired over tea at the Monomotapa Hotel in Harare, the Zimbabwe project eventually spawned forty small African-owned solar companies, which together managed to install over eleven thousand solar home-lighting systems around the country, financed by local banks. At a solar conference at the Harare Sheraton in 1995, Robert Mugabe, Zimbabwe's only leader since independence, gave the most powerful speech I had ever heard in support of using solar power to electrify the country's rural population. How sad it has been to watch Zimbabwe later implode under Mugabe's subsequent monstrous and demented rule, now one of the most corrupt dictatorships on the continent.

After speaking the same year at a solar energy conference in Swaziland, where women carried car batteries on their heads to market for recharging, I made plans to electrify an off-grid community in South Africa's KwaZulu-Natal Province. Mapaphethe was a Zulu village of a thousand households not twenty minutes from the Durban-Pietermaritzburg freeway, but it was outside Durban Electricity's reach and not in the rural electrification plan of Eskom, South Africa's huge power utility. Its young chief, who as a Zulu "Nkosi" held the power of life and death over his subjects, was distressed that he could not get his community electrified, even though high-tension power lines were visible in the distance. With the help of the KwaZulu Finance Corporation, which was willing to make bank loans to these rural people, most of whom had city jobs and a modest income, we were able to electrify one hundred fifty homes as an example of how people could have electricity *now* instead of waiting for the government-owned utility to bring it to them.

Mapaphethe became a national "exemplar" project and was visited many times by the energy ministers of South Africa and surrounding countries. The grant money was provided by the U.S. Department of Energy. The entire project cost $45,000, which covered $15,000 in expenses, including salaries for technicians, a local training course, and a loan guarantee for the finance company. The funds remained on deposit for years, earning interest that helped cover maintenance expenses. People bought their own systems at cost, supplied by Siemens Solar and installed by village technicians. This was the perfect solution: use donor funds for bank guarantees so that local lenders could offer loans for a consumer product that they normally wouldn't lend money on. Many years later, the government embarked on an aggressive and hugely expensive national rural-grid-electrification program, to the credit of Eskom and the ANC government, but solar continued to play a role. The government managed solar electrification programs of its own to reach the most remote settlements, where grid extension was too expensive.

I was happy to be back in South Africa. Since Nelson Mandela had been elected president it was a much happier place—people, as did people in the rest of Africa, now smiled at you, which they had seldom done in South Africa under apartheid.

Our solar customers smiled even more, and I never ceased being amazed when I was invited inside their earthen rondavels (round huts) to find that nearly everyone had a stereo system with large amplifiers and huge speakers, all running on solar power. Was this poverty? They had electricity and enough income to meet their needs, for the most part, which they supplemented with lush vegetable gardens, and they enjoyed the most beautiful climate on earth, with views to die for. I'd seen worse.

In 1993 I followed up on an invitation to visit the Gansu Natural Energy Research Institute (GNERI), which was western China's leading research institution at the time and fostered all kinds of solar energy. Under the direction of Professor Wang Anhua, a true solar pioneer, GNERI had developed parabolic solar cookers, solar water heaters, and large photovoltaic systems and had instituted passive solar design for government buildings. GNERI's works were scattered from Tibet to Sichuan to Chinese Mongolia, and several hundred thousand solar cookers dotted the villages of Gansu, China's second-poorest province, where they glinted in the sun like rhinestones against the brown landscape. They could boil a kettle of water in sixty seconds with the sun's focused light; fortunately, the sun moved across the sky a degree or two before the reflector could melt the kettle.

There were solar water heaters on the roofs of nearly every house in thousands of villages in China's cold northwest. Remote forest-ranger offices and road-crew barracks now had electric lights and

TV sets, thanks to Professor Wang's solar designs and relentless promotion of GNERI's solar technologies. GNERI sold most of its PV systems to government agencies for use in these various outposts, and Professor Wang installed them.

SELF hired the professor to install one hundred experimental solar home-lighting systems manufactured by GNERI in MaGiaCha, a remote village in Tongwei County. I was the first Westerner the people there had ever seen, and I had to arrange special permits to travel along this section of the Silk Road, which may not have been visited by a white person since Marco Polo (I'm not exaggerating). Mao suits were still common attire in Tongwei County, except for one young peasant girl who dressed up in a bright red jacket and donned high heels for a photograph. On my first trip to China in 1979, I visited communes where the peasants owned little more than their rice bowls. By 1993, even in this poorest of regions, where people had starved to death during the Great Leap Forward (a failed campaign to industrialize the country in the late 1950s), nearly every household managed to buy a television set once it had acquired a solarelectric system.

With funds from the Rockefeller Foundation, we launched the Gansu PV Company to manufacture the small household "plug and play" solarelectric systems that mystified visiting electrical engineers because of their amazing efficiency and reliability. Professor Wang and his son Wang Yu designed all the electronics themselves. SELF was now a 49 percent owner of the first "Sino-American" joint-venture solar company in China. And MaGiaCha became famous as the first all-solar village in the country. It was featured on TV and in the papers, and all the Communist Party officials, from the provincial vice governor on down to those at the county and district level, threw great banquets in SELF's honor and made their many pilgrimages to the remote commune nestled in the spectacular loess hills of Gansu. (More on China follows in chapter 9.)

We were "learning by doing," which became SELF's mantra, along with Goethe's dictate that action begets action. The philanthropic-foundation community seemed to agree, since the grants kept coming, project by project, as well as funds for overhead in Washington, and I kept traveling and organizing more projects. The Rockefeller Foundation, the Rockefeller Brothers Fund, and anonymous members of the Rockefeller family became big supporters of SELF's activities. I loved taking old oil money to develop carbon-free sources of power and light that didn't require reliance on fossil fuels.

The most important lesson during these years was our discovery that, in general, rural people are excellent credit risks and will do anything to make their monthly installment or bank-loan payment so they don't lose the electricity they have come to rely on. They also hate debt. For exactly the same reasons, the electric-power-cooperative movement in America was able to finance itself in the 1930s.

Back in D.C., I hired Robert Freling, an energetic global environmental gadabout. Bob had read an article about SELF in *PV News* and wrote me a heartfelt letter from Taiwan, where he was living at the time. We arranged to meet in Hong Kong, then had a follow-up meeting in Portland, Oregon, and I hired him. Bob was a Yale graduate in Russian language studies who also spoke fluent Chinese and could converse in Spanish, French, Portuguese (his favorite language), and a little Indonesian. Bob would become invaluable in China and would eventually become executive director of SELF, which he runs to this day. Two other young assistants came aboard along with several consultants. Now, we could manage all the projects I'd started and launch even more.

The next stop was Vietnam. It would be my first visit to that country since covering the war as a correspondent twenty-four years earlier. Would the Vietnamese welcome an American coming to "do good" after our war machine had killed nearly a million of their citizens— soldiers and civilians—defoliated and poisoned their forests, and dropped more bombs on their country, north and south, than were dropped on all fronts by all sides in all of World War II? I was about to find out.

3

From the Air Vietnam Airbus you could still see the bomb craters that dotted this land of green paddy fields. America actually dropped three times more bombs on South Vietnam, our ally, than on the north; even here in the green Red River delta near Hanoi, the scars were still visible. Farmers tried to fill them in, used them as ready-made fish ponds, or just left them alone.

The war had ended two decades earlier, and by 1993 half the population of this lush, friendly, somewhat laid-back country was under twenty-five years old. Young people had little knowledge of the war. Because the Vietnamese, for some reason, *loved* Americans, there would always be a strange bond between our nations. Forgive and forget was their motto, while we continued to rub salt into the wounds of our first lost war.

I wasn't here to make amends for our war crimes against this harmless land, which never threatened the United States in any way— and never could have. I was here to meet with the president of the Vietnam Women's Union (VWU), who was also the second-highest-ranking woman in the politburo in Hanoi, after Madame Nguyen Thi Binh, who had been a senior official of the Vietcong. (Note that the Vietnamese still use the French title "Madame" and its abbreviation, "Mme"—the country was once a French Colony.)

I was here because 70 percent of Vietnam's rural people, at the time, had no electricity—some six million households.

Madame My Hoa greeted me warily in her formal receiving room at the five-story mirrored glass headquarters of the VWU in downtown Hanoi. Short and plump, she exuded power and importance. We were the same age, it turned out. She was from the south and had fought the Americans in the Mekong Delta. Now, she headed up the largest semi-independent organization in the country, a huge, highly centralized social service bureaucracy representing eleven million women.

In Vietnam, women hold a great deal of power within the family and handle the family's finances. As a result, since Ho Chi Minh granted them equal rights in 1945, they also hold a lot of power in local government and are strong players in business as Vietnam heads down the capitalist road. However, only two women, Madame My Hoa and Madame Binh, were serving in the highest echelons of the central government in the early 1990s.

I explained to Madame My Hoa how SELF worked: we provided "solar seed" funds to buy one hundred or two hundred solar home systems, which we would install in households that signed an agreement to pay 10 percent down and monthly installments over three to five years. We then advised our project partner how to set up and manage a revolving fund. The partner was expected to collect the money, which would be used for the purchase of additional systems (which SELF would procure for them at cost) and to pay for ongoing maintenance. SELF would not be paid back; this was a grant. SELF would also cover its own project overhead, and it was expected the VWU would do the same (which it managed to do successfully by attracting other donors, so as not to dip into the loan fund). If the pilot succeeded, SELF would seek to raise additional funds to purchase even more systems, with the hope that the project would eventually become SELF-sufficient.

Although SELF subsidized the solar systems to varying degrees by not tacking on a margin of profit, the basic modus operandi of our approach was to get people to pay for the systems. We didn't believe in giving them away. In a country where people expected to get electricity for free from the government, it would be an achievement to get them to pay anything at all for power. At the same time, if we did not collect the full cost of hardware plus an operating margin, or did not charge sufficient commercial interest rates on the solar loans, we would be sowing our "solar seed" project's own destruction. Revolving funds will cease revolving if sufficient interest is not charged. As I have said, we were learning by doing.

In 1994 I returned to Vietnam, the second of many trips to the country throughout the 1990s, to sign an agreement with the VWU, which turned out to be the best partner SELF ever had. The women took to the nonprofit solar business like old hands, flawlessly organizing our joint solar-electrification projects in remote Mekong Delta hamlets and elsewhere in central and north Vietnam. Officials in the Ministry of Industry, in charge of power and energy for Vietnam, began inquiring why SELF wasn't working with *them*. "Rural electrification isn't the job of the Women's Union; it's our job," they told both Madame My Hoa and me.

Once the VWU agreed to be our partner, I was able to raise funds from Robert Wallace, who ran a family foundation in Washington called the Wallace Global Fund. The family fortune came from a patent on hybrid corn, and Bob had long been interested in agricultural issues in the developing world. He was the son of Henry Wallace, President Franklin Roosevelt's vice president, former secretary of agriculture, and Progressive Party candidate for president in 1948, and as such he perpetuated a family tradition of progressive thinking and activism. Normally, foundations make their decisions in secret and rarely interview the applicants for grants, but Bob wanted to see me. I'd asked for $100,000 for the Vietnam project. "I don't want to just do projects," he told me. "I want to change policy at the national level." I said I thought our project could do that, since we were already affecting policy in a number of countries.

We got the grant, and Bob came through with several more. His death in 2003 was a great loss to the environmental movement.

SELF soon acquired the support of the Rockefeller Brothers Fund (RBF—not to be confused with the much older, much larger, and more famous Rockefeller Foundation). I never dreamed I would get a hearing at this elite philanthropic institution, presided over by David Rockefeller, former chief of Chase Manhattan Bank. They read the material I sent and agreed to a meeting in New York, and again I made my pitch on behalf of two billion people living without electric light, showed photographs, and presented documents and reports. The foundation took a keen interest in what we were doing in Vietnam, and I soon received a letter saying they had approved the first of several grants for the SELF/Vietnam Women's Union project in Vietnam (see chapter 8). Every $50,000 check that fell out of an envelope, and which I rushed down to the bank to deposit, meant another 125 families would be able to purchase solarelectric systems. And the money would replenish itself, revolve several times, and finance more systems.

Perhaps SELF's biggest attraction was that we were not giving away solar electricity but were putting in place, at the local level, a SELF-reliant means for rural people to literally take power into their own hands. I believed, perhaps naively, that people in these countries should be given a chance to lift themselves up through their own efforts. Acquiring their own source of household power for lights, TV, and radio, instead of passively waiting for the government to bring it to them, was a start. Solar technology, which had become affordable for these people in the early 1990s, could do that. This message appealed to grant makers, international NGOs, local development organizations, and community groups. I even sold the idea to the World Bank, the ramifications of which I did not yet understand.

By the mid-1990s, in my working trips to nearly a dozen developing countries, I had to learn how to navigate the shifting and mysterious seas of "international development." Who did what? How did they do it? Who paid for it? I knew something of the UNDP from my earlier experience in Zimbabwe, but I didn't understand the World Bank, so I set out to learn.

In response to environmental concerns expressed by Washington's legion of environmental activists, the bank was taking its first look at renewable energy, which today, nearly twenty years later, is a huge and critical portfolio. The bank knew the UN and various bilateral agencies had been funding solar water-pumping systems for years in the Sahel (North Africa), along with the misconceived centralized PV-generating plants that I mentioned in chapter 2.

I had been hounding everyone I could meet at the bank's various departments—environment, energy, industry, country desks—to consider the idea of "solar rural electrification." I explained we were already doing it, but only on a pilot basis. Unless governments got behind it, with favorable policies and money, it wasn't going to solve the rural electricity problem, even though it was the "least cost" solution and didn't produce greenhouse-gas emissions. Most governments were in thrall to the World Bank on electric-power issues because they needed to borrow the bank's money for energy development. And the bank officials I met made it clear they were not going to fund any more conventional rural-electrification programs anywhere because it wasn't "economical." SELF was offering an economic alternative—solar electricity.

Despite SELF's lobbying efforts at the World Bank, SELF was not a policy-advocacy group. Our strategy was to set up "replicable" projects in as many countries as possible, prove the concept, and attract more private philanthropy to underwrite our very focused activity. Successful pilot projects would speak for themselves far better than our lobbying and advocacy could. We were not waiting around for the World Bank, or even governments, to fund our efforts. In any case, the World Bank only finances governments, not NGOs. I did not approach the World Bank with the idea of convincing it to stop funding big hydroelectric dams and fossil-fueled power plants in favor of projects that didn't damage the environment. I went downtown to see the bank because *that's where the money was*, and because governments needed money to pursue any sort of energy development in their countries. After SELF and other solar activists succeeded in getting the bank to open its energy lending to renewable energy, we became an outside intermediary that linked the bank's lending policies to requests for alternative energy programs by national governments and commercial enterprises.

I almost always got a polite hearing, especially from the staffers who were setting up a new department, the Asia Alternative Energy Unit (its original name at the time). It wasn't long before I was heading a World Bank mission to Sri Lanka with Dr. Anil Cabraal, himself Sri Lankan–born and the World Bank's leading expert on photovoltaics, along with a couple of specialists in electricity and finance.

I was somewhat conflicted about the Sri Lanka mission in 1993 inasmuch as I was being paid by the World Bank to do an "assessment and evaluation" of SELF's solar rural-electrification projects in Sri Lanka, which I had set up. But the bank didn't seem to mind, and we needed the bank, which had more money than all the foundations in America combined, and which allocated much of it to finance energy projects in the developing world. We wanted them to divert a wee bit to clean, small-is-beautiful energy solutions and to Dr. Ari's "people-centered development."

I took World Bank mission members around the country to meet Sarvodaya people, to see actual solar installations, and to meet happy householders who would invite us in for the customary round of tea and coconut juice. I arranged meetings with the key ministers, as well as with the Ceylon Electricity Board (CEB) and three national-development banks. I finally convinced the bank's new energy group—by then called the Solar Initiative—to finance the purchase of SHSs in Sri Lanka through these nationally chartered development banks and sundry microfinance institutions—in other words, Sarvodaya's SEEDS—instead of through the Ceylon Electricity Board.

This was a new twist for the bank, as it was accustomed to financing large-scale generating plants directly, usually in partnership with the national electric utility. It had never financed banks to make solar loans. But after five years and hundreds of thousands of dollars' worth of consulting "missions," the World Bank concluded a $32 million deal with the government of Sri Lanka and, ultimately, Sarvodaya, and the project began. At last, there was enough money to make a difference, even if it was only an infinitesimal fraction of the bank's annual energy-loan portfolio.

By the time the first money flowed, the numerous World Bank consultants assigned to the program had never heard of SELF or the story of how Sarvodaya and SoLanka, a solar NGO I cofounded, got into the solar business. The World Bank now took credit for everything, but *to* their credit, they launched one of their largest country-wide solar programs (see chapter 7 for the full story of solar in Sri Lanka).

Meanwhile, back in Washington, the World Bank was celebrating its fiftieth year, while environmentalists launched the Fifty Years Is Enough campaign, suggesting that the best gift the bank could give the world was to go out of business.

During the previous fifty years, the gap between rich and poor had grown dramatically, and the two-thirds world had gone deeply into debt on a borrowing binge that lined the pockets of corrupt ruling elites and kept the bank's seven thousand highly paid economists gainfully employed and traveling happily on their "missions." Many of these economists worked on "structural adjustment programs," which involved convincing governments to cut social programs and development projects while opening their borders to foreign capital so that investors could exploit cheap local labor for export production; thus, the subsequently "adjusted" country could pay back its suffocating debt. As Paolo Lugari, the legendary urban environmentalist once said, "Who creates poverty? Economists!"

A few years earlier, Bruce Rich had published *Mortgaging the Earth: The World Bank, Environmental Impoverishment, and the Crisis of Development,* which took the bank to task in an excoriating, heavily researched diatribe that is still worth reading today. Conservatives as well as progressives were agitating for change. Ian Vásquez, an economist at the Cato Institute and coeditor of *Perpetuating Poverty,* said, "the Bank is not reformable. It should be abolished altogether." *The Washington Post,* in 2004, paraphrased bank critics, who were "painting a grim picture of a multinational bureaucracy run amok, a secretive, bloated institution run by overpaid, jargon-babbling technocrats more interested in defending turf than tending to the poor."

But like death and taxes, the World Bank is still there, accountable to no one and no government. Its palace of stainless steel, marble, and glass piercing Washington's low skyline and dominating Pennsylvania Avenue, the bank sits unscathed by critics' attempts to cut off its funding or by the massive antiglobalization demonstrations in 2000 during the bank's annual spring meeting, which turned

Washington into a police state: all downtown streets were closed for three days so global finance ministers and bank officials could meet in peace. This was a year after the huge WTO demonstrations in Seattle. I sat with my wife in the bright sun on the White House ellipse, listening to speeches by Ralph Nader, the president of the AFL-CIO, and dozens of angry activists, and thought for a moment that I was back at a 1960s antiwar demonstration. We watched the committed young marchers set off from the rally to circle the bank (outside the closed security perimeter and under the watch of ten thousand D.C. cops and National Guard troops). I was especially delighted to see a parade float called Structural Adjustment, which featured a monstrous, papier-mâché, Godzilla-like villain gobbling up small countries.

Yet SELF had to work with the bank since it provided most of the energy financing for the developing world. The countries where we worked appealed to the World Bank for heavily subsidized low-interest loans for power "infrastructure." We wanted the same subsidies for "solar infrastructure."

In 1999 James Wolfensohn, World Bank president, personally hosted a private meeting at the bank headquarters in Washington for all the major players in the global solar energy industry. I met him there for the first time, and he promised that when he was in Vietnam the following week, he would fly over one of our solar villages in his helicopter. Wolfensohn, a likable, responsive leader by any measure, was trying to turn the World Bank elephant around and face the solar challenge head-on (which is why he called the high-level meeting on PV). His enlightened successor, Robert Zoellick, would also face an intransigent bureaucracy, but, at long last, by the end of the first decade of the new century, the World Bank began taking renewable energy projects seriously, pumping in billions to solar and wind projects around the world. According to Anil Cabraal, the bank's leading solar advocate since the early 1980s, some 2.6 million solar PV systems had been installed through various World Bank programs by 2009, totaling 138 megawatts. I like to think our prodding of the elephant in those early years made a difference.

These were busy times for SELF. In 1992 I was contacted by the Council for Renewable Energy (CRE) in Kathmandu, Nepal. This group of urban professionals, businessmen, and electrical engineers who had mostly trained in Russia during the cold war had somehow gotten hold of SELF's fax number.

Nepal is one of the poorest countries in the world. Its 23 million people live in a largely roadless, mountainous landscape with little contact with the outside world and an average per capita income of two hundred dollars. So many tourists have trekked up the Khumbu Valley to Namche Bazaar and the base of Everest that the popular image of Nepal is a place of Buddhist monks and stupas and prayer flags. In fact, this image characterizes only this one valley, populated by Sherpa people who came over from Tibet centuries earlier. Sherpas make up less than 2 percent of the population. Nepalese are Hindu, and Nepal is the only official Hindu state. (India is 85 percent Hindu, but it is a secular democracy and has no official religion.) Democracy came late to this faraway Hindu kingdom, if it ever came at all. The royal family continued to hold great power until most of them were wiped out by the crown prince in a burst of automatic weapons fire during a gathering at the royal palace in 2001. That horrific event was followed by a decade-long Maoist insurgency.

But our story precedes all that. The real disasters that beset this former paradise were climate change, deforestation, and too many people for the fragile land to sustain. From the plane, as it flew at twenty thousand feet, I could see mile-wide slashes of red earth, where heavier than normal monsoons had turned once-verdant hillsides to slurries of erosion. Meanwhile, the capital was filled with hundreds of international NGOs and their foreign staffs, who were trying to save the country from itself. This was many years before warming mountain temperatures threatened the Himalayan glaciers upon which the great rivers of the subcontinent depend.

I booked into the Shangri-La Hotel in Kathmandu after passing the cremation ghats along the river on the way from the airport. Human bodies were being ceremoniously burned, the smoke pouring over the roadway. The hotel's proprietor was a member of CRE, and he offered me a 50 percent discount and thanked me for coming to Nepal. I was soon met, fifteen minutes early, by a whole crowd of Nepalese, some in business attire, others in traditional garb, most in colorful wool sweaters and slacks. It seemed I was late for every meeting thereafter by fifteen minutes, which I discovered was because Nepal runs its clocks fifteen minutes *ahead* of the local time zone for reasons known only to the country's astrologers. (India runs its clocks half an hour out of sync with its time zone.)

Tej Gauchan, leader of the delegation, reached out eagerly to shake my hand and welcome me to Nepal. Speaking in perfect English, he proceeded to explain how this professional group came together to spur the government to bring solar power to rural people, but they had no money to do a pilot project. I had, by fax, told them to find three potential sites and pick the best one, and we'd all go visit it when I got there.

Tej said they were also challenging the "hydro guys" in the government, who dominated the debate about renewable energy, even cutting off funds to study the possibilities for solar electrification. It was obvious why the hydro guys had such power. Nepal is mountainous, and lots of water melts down from the Himalayas, so there is huge hydroelectric potential. But there is also a problem: half the population of the country lives in the Terai, part of the northern India plain, which is pancake flat and thus has no hydro potential at all, and half the mountain people live on high ridges, not in dark valleys that have little tillable land and can be flooded. For those on the mountain ridges and upland meadows, basking in sunshine, the government built small hydroelectric plants that had to be located thousands of feet below. These hydrostations required long transmission lines and year-round engineering maintenance because they were often flooded, full of silt, or sitting idle during summer droughts.

"But the water guys don't want to hear this. Nor do the commission agents, who have made great fortunes bringing in donor money to the government for hydroprojects, small and large. Hydro is where the money is," said Tej. "We need to use solar. Can you help us do it?" I told him that with all these highly trained electrical engineers committed to rural development working together with SELF, indeed I could.

Tej was one of the most interesting of the many characters I met on my twelve-year international solar journey. Born and raised in the remote western Himalayan provincial capital of Jomsom, he was from the kingdom of Mustang. Somehow, he made his way to Kathmandu, got himself an education, learned English, got married, and joined the Nepal air force, where he learned to fly. When the country decided in the 1970s it needed a national airline, he was sent to United Airlines' flight-training school in Denver, which trains international commercial airline pilots. When Royal Nepal Airlines bought its first jet, a Boeing 727, it asked Tej and two other Nepalese crewmen to go to Seattle and pick it up. He related over drinks one night at the Rumdoodle Bar in Thamel how close they came to running out of gas on the flight to Kathmandu. "We had ten minutes of fuel left. If we had missed the airport on the first pass, as many do, we couldn't have come around again." He laughed.

This highly educated and elite group of CRE volunteers, which included members of the Shahs (a royal family), high officials in the telecommunications ministry (which used PV to power remote mountaintop communication facilities), and other well-placed, middle-aged professionals, seemed out of step with this exotic, very poor, ancient Hindu kingdom. They knew more about solar energy and PV, specifically, than I could ever know. But, like so many NGOs and well-meaning people in the two-thirds world, they just couldn't figure out how to organize and manage an actual project. And they didn't have any money.

Thanks to the Moriah Fund in Washington, I was assured enough money would be forthcoming to electrify one experimental village in Nepal—its first. I never wanted to visit a remote site, appear as the white Westerner with money, and then have to let them down, but it was difficult to raise funds unless I'd already prepared a real project on the ground. This was always a tricky issue as we put together SELF's projects, and Tej drove the danger home to me by recommending a book.

Before we set off to the former Kingdom of Gorkha, home of the

Gurungs (also known as Gurkhas), in west-central Nepal to see CRE's selected site, Tej suggested I read *The Lords of Poverty: The Power, Prestige, and Corruption of the International Aid Business,* by Graham Hancock, a former journalist for *The Times* of London. It was available in every Kathmandu bookshop (Kathmandu has some of the best and most interesting English-language bookshops in the world), so I ran out and got a copy.

A more devastating indictment of the international-aid business, what Hancock calls "Development Inc.," has never been published. His exposé was worth at least a ton of World Bank, USAID, and NGO project-assessment reports, monitoring and review documents, and development studies. Hancock attacked the "aid lobby" and the colossal waste inherent in foreign aid budgeted by Western governments and the World Bank:

> The ugly reality is that most poor people in most poor countries most of the time *never* receive or even make contact with aid in any tangible shape or form. . . . After the multibillion-dollar "financial flows" involved have been shaken through the sieve of over-priced and irrelevant goods that must be bought in the donor countries, filtered again in the deep pockets of hundreds of thousands of foreign experts and aid agency staff, skimmed off by dishonest commission agents, and stolen by corrupt Ministers and Presidents, there is really very little left to go around.

What is left, he writes, is then used "thoughtlessly, or maliciously, or irresponsibly by those in power—who have no mandate from the poor, who do not consult with them and who are utterly indifferent to their fate. Small wonder, then, that the effects of aid are so often vicious and destructive for the most vulnerable members of human society."

In a country like Nepal, which has been called "overadvised and under nourished," he points out that for the price of just one foreign advisor, thirty to fifty Nepalese experts could be hired. Bringing to my attention this powerful indictment of how aid donors operated around the world was Tej's way of subtly pointing out how much foreign "experts" were hated in Nepal. Clearly, the watchword was humility. And listening.

We set off from Kathmandu in two vehicles on a cool, foggy March morning; climbed out of the high valley (which is actually a thousand feet lower than Denver); and headed down the hairpins that descend to the paved east-west highway, which leads along the spectacular Trisuli River valley to Pokhara, Nepal's second city. SELF split the cost of renting one Japanese van and one Landcruiser, plus drivers. Neither CRE nor SELF could ever afford the Land Cruisers or Pajeros that were the birthright of every aid agency and government official in every country I visited.

Turning off the paved Privthi Highway, built entirely with foreign aid, which is the only road link between India and Kathmandu, we traversed rough gravel roads through lush valleys. We stopped in small towns for tea and to pay our respects to district officials, then continued along riverbeds and dirt tracks past paddy fields and water buffalos and scattered villages with brightly colored thatched houses until we ran out of road altogether.

"Where are we going?" I kept asking, having been told little about our destination other than it was the finalist of the three villages CRE had reconnoitered for our solar project.

"Up there," said Tej, pointing to what appeared to be a human habitation stuck to the side of a hillside two thousand feet straight above us. The van became hopelessly stuck in the riverbed.

"But this road doesn't go up there, does it?"

"No, we walk."

We all had packs, and CRE had brought food and supplies and a solar demo kit. As we approached the steep mountainside rising straight out of the fluvial bottomland, fifty or so villagers emerged from the forest, smiling and waving, led by two tall and serious men named Dak and Tek Bahadur Gurung. They were the elders of the small village at the rim of the valley and of Pulimarang, the sonorously named and spectacularly sited village high above. The men were weathered and strong and dignified, talked little, and signaled for the small boys to grab our gear to carry. It was hard to guess their ages, but given that life expectancy in Nepal was forty-seven, I was sure Tek and Dak were much younger than I was.

We walked upward for two hours through the tall forests, which they told me they were carefully preserving. They understood deforestation was destroying Nepal, and they wouldn't allow it here. They also told me, via interpreters, how they had formed a village development committee that would oversee the solar lighting project that CRE had proposed to them. The people had been waiting twenty years for the government to run a power line to their village (we'd left the last power line many miles back) or to put in a hydrostation somewhere below, but it never did. They couldn't believe they might actually get electricity, and from the sun!

I asked how long people had lived way up here on this remote mountainside. "About thirty-six generations," I was told. At five generations to a century, that would be about seven hundred years. And life hadn't changed much from what it must have been like in the fourteenth century—except that there were more Gurungs.

We emerged into carefully terraced paddies worked by buffalo that appeared to have been crossbred with mountain goats. It was twilight, and I soon saw just how dark a village without power can be. Small lanterns were illuminating the large two-story stuccoed and thatched houses, and people came out on their polished stone verandas to watch the procession. My knapsack was somewhere in the throng of excited small boys. We navigated the narrow pathways by flashlight, were taken to the second floor of an enormous two-story house facing a small village square, and were bedded down by our flashlight-wielding hosts, who supplied us with piles of thick wool blankets.

"This house belongs to the colonel," said Tej.

"Who's the colonel?" I asked.

"You'll meet him later."

Some of the older CRE volunteers said they were having trouble breathing in the mountain altitude, which I couldn't understand. Pulimarang, like Kathmandu, was only four thousand feet above sea

Villagers welcome European Union delegation to Pulimarang Village in Nepal; Wolfgang Palz, front row, second from right; author, center; Dak Bahadur far left standing. *(Petra Schweizer)*

level. In Colorado, I'd lived at nine thousand feet for ten years. "Wait until you see Annapurna in the morning," said Tej.

In the morning I headed for the "living" village toilet, which was so named because it was filled with millions of wriggling maggoty worms that consumed whatever arrived from above. I emerged in the dawn light, stood on a stone wall that enclosed the path to the flimsy outhouse, looked across the cloud-filled valley, and there was the great Annapurna massif itself, gleaming in the morning sun. It appeared to be only slightly above the elevation of Pulimarang, but at eight thousand meters its summit was actually *four vertical miles* higher than where I was standing. Such is the scale of these mountains that they are beyond sensory comprehension.

I was given the full tour of all one hundred houses and farms perched on the green uplands of sun-dappled "Puli" and imagined this was how author James Hilton pictured his Shangri-La in *Lost*

Horizon. I saw an unspoiled agrarian community with no modern amenities, where the women all wore colorful embroidery and the children smiled and giggled as if they'd never seen a white man before. They hadn't. "Do they get many visitors?" I asked, knowing the international popularity of the Annapurna Conservation area, which we could see in all its awesome grandeur across the valley.

"Puli is not on a trekking route," said Tej. I asked again if any outsiders ever came here.

"Well, an English army officer came here about fifteen years ago," I was told by Dak. "That's the only white man who has ever come to our village."

The officer had been looking after the affairs of the British Gurkhas who had returned home to retire. These were the villages that had provided thousands of Gurkha fighters for the British Empire, first to defend the Raj; then to be sent to the trenches of World War I, where fifteen thousand perished; and then to the battlefields of World War II. They had last fought in the Falklands, some of the local ex-captains told me, speaking through our many CRE interpreters from Kathmandu, who found these Gurungs nearly as exotic as I did. Tek Bahadur had served in Hong Kong, where the last Gurkha garrison was stationed. Soon there would be no more work for them, and the military remittances, which allowed these villages to prosper, would end. However, retirement payments continued to arrive at the ex-soldiers' bank accounts down in the valley towns. It was these funds, the Village Development Committee informed me, that would allow at least sixty-five of the one hundred families here to purchase SHSs. Several had already bought black-and-white TVs in anticipation.

I asked how they were going to get the solarelectric systems, with their heavy batteries, up the mountain. "Just like you got here. Just like everything gets here. Walk." The teenage boys would be the porters.

A community-wide meeting was held in the village square, and Tek and Dak selected eager volunteers for the Solar Lighting Committee. Everyone assembled, about two hundred people, with all the women and children on one side of the square and all the men on the other. We took a photo of the watershed event. "I better deliver on this," I said to myself. Before the meeting broke up, the CRE volunteers and I were asked to line up. The village women came toward us in a procession, each bearing an exquisite lei of brightly colored wildflowers with which they garlanded us until we all were nearly suffocated beneath the offering. Twenty wildflower garlands around the neck are heavy! We were also given the ceremonial red-ochre third eye on the forehead, an intrinsic part of Hindu culture. That evening we were treated to a dance exhibition illuminated by the one bright demonstration solar light that we had brought.

Back in Kathmandu, we got down to work, calculating overhead costs, down payments, interest rates, and revolving-fund structure and management and preparing the technical specifics. These were achingly honest people, willing to match SELF's contribution in kind with their volunteer work. Besides CRE, my other partner was the newly formed Solar Electricity Company (SEC) of Kathmandu, started by two Tribhuvan University professors and one tough local businessman who was the Siemens Solar distributor supplying PV for the telecommunications market. They had already learned how to make their own light fixtures, inverter ballasts, and charge controllers in their little workshop. I advised them on how to improve the "user friendliness" of their electronics, based on SELF's experience in other places, and came to trust their electrical engineering knowledge and capabilities. I negotiated a contract for sixty-five solar lighting systems for Pulimarang, which would grow in a year or so to one hundred, using income from the village's revolving fund. Even Gurkha families with British army pensions could only afford an SHS with a three-year loan. SELF would purchase the 35-watt solar modules from Siemens in Singapore and airfreight them in. We decided to use solar batteries from Taiwan. CRE and SEC would work together on the installation and long-term maintenance, some of which would be paid for by SELF, and the rest covered by the solar fund. The Village Development Committee was authorized to open a solar-project account at the local bank down in the valley.

Before leaving Nepal, I took the famous sightseeing flight over Mount Everest to celebrate my fiftieth birthday. When I got back to the James Hilton Suite at the Shangri-La Hotel, there was a chocolate cake waiting for me that said "Happy 50th!" in white icing. Who knew it was my birthday? I hadn't told anyone in CRE. I called my wife in Washington, but she hadn't ordered it. I invited Tej over to help me eat it, and we celebrated the launch of the first solar village in Nepal.

Back in Washington, I followed the progress of the project via fax machine and an occasional airmail letter. I wired grant funds to local banks as needed. Citibank somehow always managed to get them through, and CRE kept track of them to the rupee. People from CRE and SEC made regular trips to check on the project, often traveling by public bus, which took about fourteen hours from Kathmandu one way, and then walking the riverbed track and up the mountain trail. All sixty-five families who had signed up had their solar systems installed by the end of the year and were making their installment payments on the $350 systems to the Village Solar Committee, managed by Tek Bahadur. This was the first credit anyone had ever extended to them.

In the spring of 1994, I received a fax asking me to come to Nepal since the country's prime minister and numerous ministers and dignitaries had agreed to officially inaugurate the Pulimarang project. However, when the date was set, I was unable to attend. Later, during a stopover in London, I was invited to meet "the colonel," at whose antiquated village dwelling we had stayed when I first went to Pulimarang. Colonel Chhatra Gurung, a veteran commander of UN peacekeeping missions around the world, was the military attaché at the Royal Nepalese Embassy in Kensington Gardens, and he asked me over for high tea. He had been the force behind the scene who had directed CRE to his home village, and he was bursting with pride at the success of the solar project, which now illuminated all one hundred families. "The lessons learned in Pulimarang are beginning to light up the lives of people throughout the whole of Nepal," he said.

"Mr. Koirala [the prime minister] flew into Pulimarang aboard his big Russian helicopter," he told me. "He brought two cabinet ministers, many officials, and many press and TV people." The prime minister had placed a gilded marble plaque in the village to commemorate the inauguration of Nepal's first solar energy project of its kind. Balancing my teacup, I examined the photographs and press clippings of the event. I had met with the prime minister earlier in Washington and had urged him to support a national program based on the Pulimarang model, and he had done just that. Through the country's network of agriculture banks, a 50 percent subsidy was offered to borrowers purchasing SHSs from any of three qualified companies that were just getting into the business: our partner, SEC; Lotus Energy, run by an intense expatriate American Buddhist; and Wisdom Light Group, owned by a Tibetan refugee. All three companies and a dozen new ones grew rapidly, selling and installing tens of thousands of SHSs all over the country.

Colonel Gurung, a kindly, thoughtful, dignified man in his mid-sixties, had accomplished, with the help of SELF and all the local

Dak Bahadur and the author in Gurkha village of Pulimarang after the installation of the first seventy-five solar home lighting systems. (*Author*)

volunteers, his lifelong goal of bringing electric light to his native village, leapfrogging his people from the fourteenth century directly into the twenty-first. Because Nepal had a geostationary communications satellite, as well as solar-powered TV-broadcast repeater stations on mountain ridges all over the country, it was possible to pick up TV signals easily in the remotest communities, if there was electricity. Colonel Gurung explained that more than half the families had saved up and bought 12-volt black-and-white TV sets. One retired Gurkha captain had used his British military remittance to buy a satellite dish and a *color* TV, becoming a fan of British soccer and the world news from the BBC.

The colonel explained that solar electricity had gotten a bad name in Nepal because four large French-government projects (based on the minigrid approach discussed earlier) had failed miserably, so he thought it might be a risk to propose his village as the guinea pig for SELF's first solar-electrification project in Nepal. Accompanying the CRE on its scouting mission, he had trekked up to Puli and asked the village elders himself what they thought. "They had never heard of solar power," he told me, "and I explained the solar systems weren't free. It was difficult persuading them of the benefits of solar power. It was hard for them to believe such a thing was possible. Now, I get reports that people come from all the surrounding villages to visit Pulimarang to see the solar lights, and they want them for their villages, too."

Over the years, academics and aid consultants came from Germany, Denmark, the United Kingdom, and Canada to study Puli as a "shining example." One young German woman lived there for several months and wrote her doctoral dissertation on the social impacts of electricity on a remote Nepalese community. We had done our job with our small grant, and people like Tej and the colonel and the professionals at CRE made it happen. These are the kind of people who would be the salvation of the country. This was SELF's model, seeking out people who cared about their own countrymen and who only needed a small amount of money and a little bit of encouragement to do more with $1,000 than USAID could do with $100,000. In the end, the entire cost of the project that launched a "solar revolution" in Nepal was equal to the duty-paid price of one imported Toyota Land Cruiser.

Five years later, I returned to Puli with a small delegation that included Dr. Wolfgang Palz, the pioneering chief of renewable energy for the European Commission, and later the EU (see chapter 10), and the vivacious German woman who'd made Puli the subject of her Ph.D. thesis. I was welcomed back effusively by Dak and Tek, who took us round the village to show us that every solar system was still working fine. Only a few batteries had been replaced, and SEC had upgraded the electronics where necessary, paid for out of the village solar fund. SELF had topped up the fund for battery replacement. Tek showed me the yellowed fund ledger, which posted each villager's installment payment until the loan was paid back after thirty-six or forty-eight months. The fund had "revolved" enough to purchase another thirty-five SHSs for the rest of the families. TV antennas sprouted from nearly every house; women told me they could weave on their porches after dark under their bright lights; men said they felt, at last, "part of the city" by having TVs and radios. Nepal TV, which had produced a half-hour documentary on Puli, provided educational programs on agriculture, literacy, and language. Once a week the development committee set up a solar-powered TV and VCR on the colonel's veranda and showed feature films in the village square. Children studied after dark, which they could not do before. "Having light makes us different from all the other villages," said the man who owned the color TV.

And again the women smothered us in flower leis. The village children had strung together marigold garlands, and each of them had to bestow his or her garland on one of us as we made our way back down the trail to our rented Land Cruisers. We draped the garlands all over the vehicles, which looked like floats in the Pasadena Rose Parade, and headed back to Kathmandu. Later, Pulimarang got its first telephone line. I was given the number of the one telephone, at the Village Development Committee office, and I thought it would be fun to call Tek or Dak, but I never did because they don't speak English and all I could say in Nepali was the traditional greeting, "*Namaste*."

* * *

SELF was now hitting its stride. Money was coming in. Our little staff worked long hours and traveled the globe. Bob Freling and I worked up plans for projects that had been requested in Indonesia, Brazil, and the Solomon Islands. The latter came about when I broke my arm hiking in the Alps, and the Swiss doctor who put my mangled wrist back together at his clinic in Meiringen was interested to learn what I did. When I told him, he invited me in my elbow cast up to his all-solar house featuring photovoltaics, solar heat, solar water heating, and passive solar design. He said, "I believe in solar energy."

Dr. Oberli then told me he was giving up his "spoiled good life" in Switzerland to move with his wife to manage the main hospital in Guadalcanal in the Solomons. "I could help you establish a solar program for the Solomons," he volunteered. I sent Bob to the Solomons, and he set up the Guadalcanal Solar Rural Electrification Agency with the help of Dr. Oberli. Soon, bright lights from the high-peaked hardwood and thatch houses could be seen twinkling along remote island shores in the night.

Children read after dark under solar lights at rural home near Bangalore, India. *(Philip Jones Griffiths)*

SELF also launched small pilot projects in Indonesia, on Java, and on Brazil's sandy, windswept northeast coast. I continued to commute to China and Vietnam, with visits to South Africa in between, and considered the idea of setting up a base in India, which the World Bank had asked us to do (see chapter 6).

At home, looking for partners and money, I visited Americus, Georgia, where Habitat for Humanity was based. Habitat's president liked the idea of joining forces, so we prepared a joint project for Uganda. I raised $150,00 from the U.S. Department of Energy and sent former DOE PV program director Paul Maycock, from SELF's board, to Uganda to set it up. But after the DOE grant had been tapped out for 150 houses, future Habitat householders got houses without electricity—our SHS added $400 to the cost of the $700 homes! Too much for poor people. But in the meantime, these widely watched and studied efforts launched a batch of private solar enterprises in Uganda.

By 2012, small, inexpensive solar lighting systems developed by the World Bank and by American and European NGOs, featuring LED lights, were lighting the homes of Africans who couldn't afford our original SHS that required larger PV panels and batteries; in the 1990s, PV was still expensive, and LEDs, which use very little power, had yet to come into the market (see following chapter).

In Africa, SELF also provided funding to electrify health clinics, schools, and community centers in a remote Masai settlement in Tanzania. Working with the young, educated tribal leadership was a delight, especially knowing they were trying to save their people from the encroachment of "civilization" by using the latest technologies available. They needed TV, radios, mobile phones, and computers to survive. Not long after, Masai warriors could be seen walking around with a spear in one hand and a cell phone in the other. (Kenya and Tanzania have better cell phone coverage than some parts of the United States.)

SELF was satisfying a need in many of us to "make a difference." Because SELF was a nonprofit organization under IRS rules and did not need to make money, we were free to experiment and to be as innovative as possible. I discovered that *anyone* can make a difference, *if they want to.* You just have to believe, get moving down the road you have chosen, and don't stop. I have always believed that the moment one definitely commits oneself, the universe moves, too.

My mantra became, and still is, the "Seven *Is*": imagination, inspiration, intuition, innovation, insight, ingenuity, and integrity. My other motto, which I was using long before Nike thought of it, was: "Just Do It!" (Well, actually, it was, "Just fucking do it!"—"JFDI.")

SELF was "doing it." One validation of SELF's success came in 1996, in a letter from the CRE in Nepal that explained how our "solar seed" at Pulimarang had grown into a national program. "A sacred fire ignited by you is spreading," Tej Gauchan had written. Now it was time to spread that fire wider and faster around the world, and eventually start a solar wildfire in the United States (that would come much later: see chapter 11).

But first, a little background about the technology you've been reading about.

4

Solar electricity may seem like a gift from the gods, but it was an invention of man (perhaps God inspired), and it is not new. In fact, it is more than a half century old. In 2004 the silicon solar cell celebrated its fiftieth anniversary. Well, the cells couldn't celebrate, but the people involved in developing the conversion of sunlight to electricity celebrated, or at least those scientists and engineers who were still alive to mark the date.

The actual discovery of the photoelectric effect dates back to the Becquerel family, nineteenth-century French physicists who first experimented with electrochemistry in 1839 and created a "voltaic cell" that produced a current. In the nineteenth century, numerous pioneers working with photochemistry in their laboratories discovered that light had an effect on certain solid materials, such as selenium, creating a flow of current. One inventor, Charles Fritts, sent his experimental "photoelectric plate" to Werner von Siemens (who Americans refer to as the German Edison, while Germans refer to Edison as the American Siemens), who proclaimed that photoelectricity was to be of great importance. In 1876 Siemens himself reported to the Berlin Academy of Sciences on light's impact on selenium's electrical conductivity. A hundred years later, Siemens, the global manufacturing giant, injected millions of dollars into its subsidiary, Siemens Solar, which for most of the 1990s was the world's largest PV company. (Siemens for some reason exited the PV business as the millennium dawned, then reentered the solar manufacturing business a few years later, then left it again in 2012.)

Initially, solar photovoltaic cells were called "solar batteries" because the voltaic and galvanic (remember the scientists Volta and

Galvani?) chemistry that stores electrical energy in conventional batteries is similar to the solid-state chemistry of the photo*voltaic* effect, in which the photons of light, hitting the right conductive material (originally selenium, then chemically treated silicon), knock electrons out of their atomic orbit, creating an electrical current. I don't mean to get too technical; I'm simply trying to rectify the language and explain how the almost unpronounceable name "photovoltaics" became attached to this wondrous technology.

It wasn't until Einstein's work on the composition of the atom in the early twentieth century provided an understanding of how subatomic particles such as photons and electrons interacted that the actual physics of the photovoltaic effect were comprehended. Einstein was awarded his only Nobel Prize for this early work of explaining what energy writer Daniel Yergin calls "the alchemy of sunlight."

New Jersey's Bell Laboratories, known as the "idea factory," developed the first actual solar cell in 1954. Three scientists at Bell chemically configured small strips of silicon that could convert 6 percent of the sunlight that struck them into electric current. These "power photo cells" were presented to the public as the "Bell solar battery." A very small panel of cells powered a twenty-one-inch Erectorset Ferris wheel. *The New York Times* made the invention a front-page story, and the solarelectric age was born.

This early history and the ongoing development of PV are thoroughly chronicled in a marvelous book called *From Space to Earth: The History of Solar Electricity*, by John Perlin. (His nineteen-page chapter, "Electrifying the Unelectrified," examines the international off-grid solar market my two organizations were developing.) There is no need to restate Perlin's concise, scholarly telling of how solar electricity came to be. I would refer you to Perlin. (I must also refer you to Wolfgang Palz's remarkable historical compilation, *Power for the World*; see chapter 10.)

The "space" part of Perlin's title refers to the first successful use, or "application," of Bell Lab's "solar battery." At $286 per watt, Bell Labs figured it would cost a homeowner about $1.5 million to power a house! Bell's Western Electric Company found a commercial application, using small PV modules to boost telephone signals on remote lines in rural Georgia. But silicon transistors came along, which could amplify voice signals at a much lower cost. Then, a real market was found: the U.S. Navy was persuaded to put a tiny solar battery on one of its *Vanguard* satellites in 1958. The first solar-powered transmitter from space worked perfectly, followed by *Telstar* in 1962 and *Skylab* in 1973. As hundreds of communications and weather satellites were launched in subsequent years, an industry was born and numerous semiconductor companies began turning out silicon solar cells. The cost of cells didn't matter, since there was no alternative power source in space.

The U.S. Air Force, Army, and Navy were all putting up satellites after the Soviets humbled us in 1957 with their *Sputnik,* the first human-made device ever rocketed into orbit. The Russians quickly matched our scientific prowess in solar cells to power their future satellites when *Sputnik*'s onboard batteries failed. Bell Labs and the many researchers working on solar cells had not kept their discoveries secret. So the solar space race was launched. Years later, I met John Thornton, one of the army physicists who had worked on the first solar-powered military communications satellites, and he told me that space-grade solar cells he had made in 1961 were still powering signals from some of those original satellites in 1996, thirty-five years later. (More amazing still, the original solar cell unveiled by Bell Labs in 1954, on display in a touring exhibit, was still producing a small current in 2004!)

One of the visionaries who helped the solar industry link space and earth was science fiction writer Sir Arthur C. Clarke (author of *2001: A Space Odyssey*), who in 1945 had published a *nonfiction* article proposing that television signals could be bounced around the world from earth to space and back to earth via solar-powered communications satellites positioned in geostationary orbits. That idea certainly looked like science fiction way back then, since few people at the time had heard of television, solar photovoltaics, satellites, or even rockets powerful enough to launch a satellite into orbit. But Clarke saw all this coming. Years later, in Sir Arthur's office in Colombo, Sri Lanka, where he lived since the 1960s until his death in 2008, he told me how happy he was to see SELF fulfilling his vision of bringing solar-powered televisions to the country's rural villages. He also lamented that he was never able to patent the communications satellite, the idea that made Marshall McLuhan's "global village" a reality. It was a technological breakthrough—made possible by the solar cell—perhaps more important than the Boeing 707 and 747.

When I first met Sir Arthur in 1992, our Sri Lankan operations were in daily contact with me in Washington via satellite telephone networks (later replaced by high-bandwidth fiber optics, something Sir Arthur also predicted). And SELF eventually set up solar-powered-satellite Internet links in remote communities in the Amazon and South Africa. Without solar PV on the satellites in space and powering the uplink transmitters, downlink receivers, and computers on the ground, the isolated residents of the Xixuaú-Xipariná Reserve in Roirama State, Brazil, and the grateful students at Myeka High School in Mapaphethe, KwaZulu-Natal, would never have joined the modern world (see chapter 9).

We had Sir Arthur to thank, and thank him I did every time I was honored to visit him at his large home in Cinnamon Gardens. I would update him on our latest endeavors worldwide, and he would show me the latest downlinks from NASA, received by the satellite dish on his roof. On his laptop, he once showed me NASA's first satellite-generated, infrared, digital photomap of the world that clearly marked, with visible light, the places where humans had electricity. In huge areas of Asia, Africa, and South America there was only darkness. "See," he would say, rolling his wheelchair away from his computer desk, "you have lots of work to do!" One of his last wishes, he e-mailed me just after his ninetieth birthday, was that "the world adopt clean energy sources." We carbon-based bipeds, as he called the human race, lost a great prophet.

Indeed, there was work to do on earth, just as work had been done in the heavens. Many of the manufacturing pioneers who had started companies to serve the government-funded space program and telecommunications industry were interested in "terrestrial applications." They wanted to bring their technology down to earth, to serve humanity, replace fossil fuels, and bring power to people beyond the electric grid.

Scientists and engineers in many countries had become so enchanted with this sublime source of power that they abandoned the scientific method, gave up their engineering principles, and began *dreaming*. They dreamed of using this technological breakthrough in ways that would have "societal impact" and that would put the "sun in the service of mankind," which was the theme of the first Solar Summit in 1973. One dreamer, Dr. Harry Tabor, a well-known research scientist, published a paper in 1967 proposing "Power for Remote Areas" using solar cells.

Another scientist-dreamer—and doer—was Dr. Elliot Berman, an industrial chemist who founded the Solar Power Corporation in 1969 with backing from Esso (Exxon). At the Solar Summit in Paris in 1973, according to author John Perlin, the company announced it had "commercialized and [was] marketing a solar module . . . which [would] compete with other power sources for earth applications." Berman wanted to provide electrical power to those in need, he told Perlin, especially "those who live in rural areas of developing countries." (I didn't know I was following in the footsteps of giants when I founded SELF.)

Berman promoted the idea that lowering the cost of manufacturing using existing silicon-wafer technology was the better strategy, rather than focusing on research to make cells more efficient, which was the approach driven by the space program, where costs didn't matter. On earth, costs mattered. While the space program would pay one hundred dollars a watt for highly efficient space-grade solar cells, Berman knew he had to bring the costs down to earth to sell on earth, and he did—to ten dollars a watt. The first market he found was the U.S. Coast Guard, which had twenty-five thousand navigational buoys that needed their batteries replaced periodically. If the Coast Guard could install batteries that would be recharged by the sun, maintenance trips out to the buoys would no longer be needed.

The men who took Berman's dream to the next step were Dr. Joseph Lindmeyer and Dr. Peter Varadi, who founded Solarex Corporation in 1973, shortly after the Solar Summit. With a pocketful of space program, military, and telecommunication orders in hand, the former communications satellite scientists built a spectacular "solar breeder" north of Washington, D.C., in Frederick, Maryland. The "breeder," the result of scientists dreaming again, was a slab-roofed factory covered with a 300-kW array of solar cells that supposedly would produce enough power to produce more solar cells. But it took far more power than the system could generate to run all the complex, high-tech fabrication equipment inside. The giant blue roof array, shimmering in the sun and visible from the interstate highway, provided backup power via large standby battery banks that kept the production line running during rare power outages. The iconic Solarex building continued to churn out solar cells and modules, night and day, for thirty-eight years, until BP Solar, which had purchased Solarex as previously mentioned, shut the factory down in 2011. The building was demolished the following year.

In its early years, Solarex quickly expanded, Dr. Varadi built more factories in Europe and Asia, and the company was soon selling solar modules by the megawatt, now the standard unit of measurement for quantities of solar-module production. When they hit 1 megawatt, they celebrated. (A million—*mega*—watts is the equivalent of 10,000 100-watt solar modules.) These "PV shipments" were in addition to the space-grade solar modules Solarex produced for NASA. A big market were "navaids," or solar-powered navigational buoys and foghorns, which were supplied to the oil companies so they could identify their myriad offshore drilling installations and rigs in the Gulf of Mexico and elsewhere.

A little clarification here: 1,000 watts is a kilowatt, 1,000 kilowatts is a megawatt, and 1,000 megawatts is a gigawatt. One kilowatt in the developing world will light up twenty-five homes. In the United States, the average American residential installation is now 5 kilowatts. A 1-megawatt PV array will provide most of the power needed by two hundred homes when coupled with energy efficiency and conservation.

Another large market was supplying solar modules to the oil and gas industry for use in "cathodic protection"—stopping corrosion on hundreds of miles of pipelines and thousands of wellheads. These solar modules sent a small electrical charge through a pipe into the ground to neutralize corroding molecules. The major oil companies became big customers for Solarex's corrosion-proofing solar solution. And many of them eventually got into the solar business themselves (and then most got out).

Joseph Lindmeyer died just after I started SELF, but one day I tracked down Peter Varadi, who was already legendary when I first consulted for Solarex. I wanted to learn more about the early days of solar PV. I had no idea where he lived, but his area code indicated Maryland. When he gave me his address on the phone, I realized he lived in the large condo building in Chevy Chase *right next door to me,* and as we determined where our apartments were located in our respective condo towers, we discovered we could lean out our windows and wave to one another! Peter became a friend and supporter of our work in the developing world. Peter had sold Solarex to Amoco (Standard Oil of Indiana) in 1983.

A chemistry graduate of a Hungarian university, where he earned his Ph.D., Dr. Varadi escaped from his homeland just as the Russians invaded to crush the Hungarian revolution in 1956. It was a gain for the United States, as he went to work for COMSAT in Maryland researching the expensive handmade solar cells that powered satellites. I asked Peter what motivated him, and he said it was the dream of using solar not just to power satellites but to bring electricity to remote communities. He told me he first saw solar modules powering mountain homes in Spain, which in turn convinced him to open a half dozen factories in Europe. Soon, Solarex became the world's largest solar company, the bulk of its 100-megawatt annual production going for all kinds of "remote power applications." Today, in retirement, Peter works on PV standards and recently finished a book on the "human side of PV."

During the 1970s, the DOE's PV program director, Paul Maycock, was very busy handing out contracts and grants to the solar PV industry to help take it beyond research and into commercial markets besides the space program. The Solar Photovoltaic Energy Research, Development, and Demonstration Act of 1978 had committed

$2 billion to this technology under President Carter. The objective of Paul's well-funded program, managed by the Jet Propulsion Laboratory in Pasadena, California, was to bring "per watt" costs down to a *dollar per watt*. (Unfortunately, this would not be achieved until 2012!) There were two ways to do this: by producing higher volumes of cells with efficient production methods or by producing more efficient solar cells. A great deal of progress was made on both fronts. Unfortunately, the program's early success was cut short when it was dismantled by President Reagan in 1981, and Paul resigned. His prediction of two-dollar-per-watt solar modules by 1990 would not be realized until 2008.

What originally saved the industry was the oil companies. When people would ask me, "Whatever happened to solar energy," and I told them, "The oil companies bought it," they immediately reached for a conspiracy theory. "The oil companies want to suppress it" is the paranoid view I've heard reiterated a thousand times. Solar pioneer Bill Yerkes started a small solar company in 1975 to manufacture solar cells for the new terrestrial markets—navaids and cathodic protection. It was quickly purchased by Atlantic Richfield (ARCO), which launched ARCO Solar. ARCO Solar was slated to be a one-billion-dollar-a-year company by 2000. Spending millions on R & D and state-of-the-art automated manufacturing, ARCO Solar brought the price down to $8.50 per watt. By the early 1980s it surpassed Exxon's Solar Power Corporation, BP Solar (British Petroleum), Mobil Solar, and Total Solar, the French oil giant's PV company. As previously mentioned, ARCO Solar was later sold to Siemens.

Without the oil companies' early money, which they invested partly because of their experience using PV to prevent pipeline corrosion and partly because of the oil crisis of the 1970s, when investing in any other source of energy seemed like a good idea, PV would have remained a niche product. As Charlie Gay, the former head of the National Renewable Energy Laboratory (NREL), former president of ARCO Solar, and cofounder of Sunpower Corporation, told me, "Companies don't invest hundreds of millions of dollars in a technology because they want to destroy it." In the late 1980s Exxon and Mobil got out of the direct operating solar business altogether, and Shell exited a decade later.

After it bought ARCO, Siemens Solar became the best of all the solarelectric manufacturers, making the finest product with the longest warranties and the fairest prices. I for one was pleased to see a real electrical manufacturing firm, second only to GE, get into the solar business. I became acquainted with Siemens Solar president Gernot Oswald, who took a deep interest in the developing-world markets I was working in. He visited remote sites of ours, tramping through swamps and across monkey bridges and into jungles. To my knowledge, no other senior executive in the global solar industry had done that. Gernot loved seeing Siemens Solar modules atop thatched huts in Vietnam, Nepal, and South Africa, changing their inhabitants' lives. He put a 6-kilowatt system on his own house in Munich. He was equally proud of the 5-megawatt solar PV roof Siemens constructed atop several large buildings at the Munich Messe (trade fair center) in 1999, the largest array of solar PV in the world at the time. He showed me around the installation with the enthusiasm of a small boy (he was in his late sixties at the time). Then, in 2001, Siemens decided to get out of the PV business and sold Siemens Solar to Shell, the world's second-largest oil company.

BP Solar and Shell Solar began competing for solar markets. "The sun holds such bright promise as a clean, renewable energy source" announced Shell's ubiquitous TV and print ads as the oil giant explained that it was working to make the dream of solar energy come true. Meanwhile, BP Solar had reportedly spent $50 million on its corporate sunburst/sunflower logo promoting "Beyond Petroleum" and was criticized in a campaign by Greenpeace International that targeted BP shareholders and pointed out BP spent more on advertising its new sunburst logo than it did scaling up PV manufacturing at the former Solarex plant in Maryland it had acquired. Both BP Solar and Shell Solar would close down worldwide operations within ten years (except for Shell's Japanese subsidiary, Showa Shell, which today is building a 350-megawatt PV factory). I could say, "Never trust an oil company," except that France's oil conglomerate, Total, soon became a major solar-project developer and PV producer after it bought America's largest PV company, Sunpower, in 2011. Never try to understand an oil company, but *do* get out of their way when they decide to move.

So exactly who was buying all this solar PV, besides the telecom industry and the military, that the oil companies were so eagerly producing? The early market here at home in the 1970s and 1980s—which no one wanted to talk about at the time—was largely in northern California and the Rocky Mountains, where marijuana growers needed off-grid power for their houses and processing operations. No grid, no electric bill, no trace. It was estimated that California's largest cash crop in the late twentieth century was marijuana, earning billions of dollars for tens of thousands of small growers scattered throughout the state. Solar retailers like Real Goods, Backwoods Solar, and others sprang up, managed by "hippies" who made a nice living selling solarelectric systems to their counterculture colleagues, who were making even more money growing weed. The executives at ARCO Solar and Solarex never inquired as to where all these shipments of solar modules were going, but lots of mom-and-pop solar dealers were paying top dollar for whatever they could get. Marijuana dealers and PV retailers formed an unwitting alliance that contributed more to bringing solar electricity down to earth than anyone wants to admit.

Solar energy appealed to the counterculture, to the disaffected, to society's rebels and social activists, and to those simply wanting independence from "the system." Solar could certainly provide that. A citizens' movement evolved to push the solar agenda politically, culminating in the 1970s with the work of the Solar Lobby in Washington, which published *Blueprint for a Solar America* in 1978.

After going back to Colorado to work on nonsolar projects during the Reagan years and then returning to Washington in 1987 to work for Greenpeace, I had been out of the loop with regard to solar energy for seven years. Greenpeace scientists were warning that if the world didn't soon reduce greenhouse-gas emissions, in twenty years the world's weather would go straight to hell and "climate change" (a new term then) would bring on extreme weather-related disasters, which we all know has come to pass since the world did nothing for two critical decades (and after Rio + 20 intends to do nothing now).

I learned enough about global warming at Greenpeace to know that the world had to kick its fossil-fuel habit, so I joined Solarex as a consultant for eighteen months, hoping to learn about possible solutions in clean energy. I researched the *other* market few people were talking about: *household solar electrification in the rural two-thirds world.* This market was booming, I discovered.

Half the shipments of solar modules to developing countries were ending up serving off-grid customers in rural areas, but it was also perceived as a problematic market since "these people don't have any money," as you'll recall I was informed time and again. However, the donor agencies did have money, and they would pay for solar installations in developing countries, while NGOs and consultants would seek to implement the vision of E. F. Schumacher, British author of the worldwide bestseller *Small Is Beautiful.* Schumacher talked about "intermediate technology," and NGOs promoted "appropriate technology."

In the African nation of Mali, solar water-pumping systems were developed as an effective life-saving business, and donor money poured in . . . rather, it poured into the pockets of the solar companies and the NGOs and consultants like the Intermediate Technology Group, which morphed into IT Power, directly inspired by Schumacher. Based in the United Kingdom, IT Power, under the longtime, tireless leadership of Bernard McNelis, did more in the early days than any other single organization to promote solar as a solution to real problems in the developing world (e.g., using solar for vaccine refrigerators and pumping, using PV for health clinics and government telecom). It is still going strong. IT Power pioneered community solar "applications" in China, all over Africa, and in South America, and it trained some of the people who later became SELF project managers and SELCO entrepreneurs.

Even though the solar manufacturers did not understand this

market in the way they understood the space, navaid, telecom, and cathodic-protection markets (and, yes, the marijuana-grower market), the PV companies nevertheless could not help noticing that off-grid electrification had become a major driver of their business. But they were not interested in serving this market directly—it was deemed just too difficult. They were right about that! The closest the solar manufacturers wanted to get to the developing-country market was the telecommunications business. It was far easier to meet with communications ministers and sell them a huge order for solar-powered mountaintop repeaters than to traipse around in the rural areas looking for individual customers. Thus, as you've seen, an opportunity and a challenge were waiting.

In the United States and Europe today, 99 percent of solar installations are "grid tied," which means they feed power directly into the utility grid after it has been "conditioned" and inverted to AC to match the sine wave and cycles of the utility power being distributed. Electric meters run backward when the sun is shining. At night, since there are no batteries in these systems, the power is drawn from the utility power lines. A quarter million American households now have rooftop solar systems to augment or replace utility power (and many more will have them by the time you read this).

If they are "stand-alone" and not connected to the grid, they use large banks of deep-cycle batteries instead. The original solar pioneers in America were people in remote houses or cabins who had no choice but to use solar electricity with large battery banks (i.e., the marijuana growers!) or "early adopters" who just wanted it, like former Republican congressman Roscoe Bartlett, who built the first solar house in Maryland nearly twenty-five years ago. A growing market will be people who want to get "off the grid" and become independent of their utility company, but it's not a practical solution for most high-energy-consumption American homes.

So what is it going to cost "me" to put in a home solar system? First, you need to know that the cost of solar PV is calculated by "per-watt installed," referring to whole systems, or by "per watt" (per Wp), referring to solar modules themselves. Prices have fallen more than

75 percent in the past few years as manufacturing capacity increased worldwide, especially in China. You can order one solar module on the Internet for under a dollar a watt. This does not include the cost of installation. Today, a residential solar system in America can be installed in most states for around $2.75 a watt, which means a 5-kilowatt (5,000 watts) home solar installation would cost less than $14,000 without adding in the 30 percent federal tax credit or any other incentives or local grants and tax credit available (there is no sales tax on home solar installations). You can't afford *not* to go solar!

For a detailed discussion of PV and other solar technologies, along with energy storage, competing energy sources, and cost calculations for solar installations in both the two-thirds world and the industrialized world, please see Appendix 1: "Solar Tech Simplified." See chapter 11 for much more about how Americans are using solar power.

5

When historians a thousand years hence look back at the twentieth century, they will see two main things that defined our time: electricity and oil. Electricity, thanks to Edison, Westinghouse, Siemens, and Tesla, changed the way we lived. Oil powered the century, replacing wood and providing fuel for transport, which coal could not do.

Before cars appeared on the scene, however, and before electricity became popular, the American oil business was built on the global demand for kerosene, a refined fuel that burned more brightly than whale oil or the vegetable, animal, and mineral oils that humans had been burning in their lamps since they lived in Mesopotamia. The largest fortune the world had ever seen was created by selling refined lamp oil under the manufactured name "kerosene," and John D. Rockefeller became the world's richest man by lighting the oil lamps of the world.

In today's electrically lit world we forget the powerful human need for lighting. It is only brought home during occasional power outages and area-wide blackouts. Everything stops, but, worse than that, it's dark! Because the sun is on the other side of the world half of every day, we spend half our lives in the dark. Or so we would without lighting. The worldwide demand for lighting is responsible for the biggest business industry of modern times—electric-power generation. It's so big we don't even see it, and we take the supply of electricity and electric light for granted.

John D. Rockefeller had only one market for his product initially: oil lamps. When kerosene was refined, the by-product, gasoline, was flushed into rivers because there were no cars to use it. Lubricating oil and grease were minor by-products. Fuel-oil furnaces to heat houses hadn't been invented. Rockefeller might be considered a "robber baron," but he was also the biggest philanthropist America had ever seen, and he did bring kerosene lighting to poor people since kerosene was cheaper and more plentiful than whale oil, which only the rich could afford. "Let the poor man have his cheap light," he liked to say.

Kerosene was a new fuel, extracted from petroleum coming out of the ground at thousands of wells in Pennsylvania, Ohio, and Indiana, where the global oil business started. Because of impurities from poor refining methods, oil lamps were always blowing up, killing and maiming people, and burning down houses. As Rockefeller consolidated the industry and created a worldwide distribution business that he controlled, he wanted his oil to be safe. He didn't want to kill people. He had standards. So he named his company . . . Standard Oil (SO). He said his oil wouldn't blow up because it was "standardized"—carefully refined according to certain safety and production standards. People knew what they were buying. He was not only the world's greatest industrialist but also its first marketing genius. Safety sells.

At the Museum of the People on Beijing's Tiananmen Square there was once a display showing how the Western oil companies in the late nineteenth and early twentieth centuries "exploited" Chinese peasants by giving away oil lamps so that they would buy kerosene. Mounted in a display case were the glass wick lamps of Standard Oil, Texaco, Gulf, and Shell. Retired oilmen I met years ago remember from their youth the campaign by SO (Esso) "to light the lamps of China." In fact, the corporate magazine of Esso, later Exxon, is still called *The Lamp*.

When I started SELF, I appealed for funding to the Rockefeller Foundation and the Rockefeller Brothers Fund (RBF), as well as to members of the Rockefeller family, on the basis that hundreds of millions of people around the world were still lighting their homes with smelly, dirty, and dangerous kerosene (not all of it so "standard"). I had a conversation with Larry Rockefeller in his office at the Natural Resources Defense Council in New York one day, and he thanked me for reminding him how much of the original family fortune came from oil for lighting.

Speaking for SELF, I said it was time to replace every nineteenth-century kerosene lamp on earth with a clean, affordable, reliable, twenty-first-century form of energy for lighting, which happened to be solar PV. I made the case that the Rockefeller philanthropies should help end the era of carbon-based lighting, which poured millions of tons of greenhouse gases into the atmosphere annually. Let's help "bring the poor man and woman their cheap solar light," I urged. It worked. A year or so later I was pleased to hear a speech by Rockefeller Foundation president Peter Goldmark at a renewable energy conference purposefully advocating solar solutions for the third world.

The only difference between our proposed commercial approach and that of JDR a century earlier was—and this is the crux of the matter—that we could not give away lamps and bottle sunlight. If we could lease the sun and sell sunlight, we'd give away the solar cells and the PV modules. But other than that detail, the market-place for our new lighting product was identical to JDR's, and it had not changed in the more than one hundred thirty years since he launched Standard Oil back in Cleveland, Ohio, in 1868. And the market remains huge today: the World Bank has estimated that $36 billion is spent globally each year on kerosene to illuminate homes.

The Rockefeller Brothers Fund called me up one day in 1995 and invited SELF to organize a small three-day workshop at their Pocantico Conference Center in Tarrytown, New York. I was stunned, honored, and grateful that they thought enough of SELF's activities to make this offer. The conference would be called Selling Solar: Financing Household Solar Energy in the Developing World.

We put together a short list of one hundred fifty invitees. Then, the foundation told me that the conference center only had twenty-six seats, positioned around a big circular table. So, we shortened the list and came up with a roster of twenty-six men and women who represented every aspect of the subject at hand: solar electrification in the developing world. Each invitee was required to commit three full days to the event. I was reminded of the similar effort to "sell solar" to the American people that Tom Tatum and I organized for the DOE back in 1979 at Gold Lake Ranch in Colorado. This event would turn out to be far more important.

To address the issue, we invited leading representatives from the World Bank, Greenpeace, the Global Environment Facility, the solar industry, the United Nations, the global reinsurance industry, Wall Street, the trade press, and the philanthropic world.

Notables included Paul Maycock, of SELF's board and publisher of the industry newsletter *PV News*; K. M. Udupa, deputy manager of Syndicate Bank in India; Dr. Charles Gay, chief of the National Renewable Energy Laboratory (NREL); Richard Hansen, of Enersol (see chapter 2); the number-two executive of the Grameen Bank of Bangladesh; and the wonderful facilitator, Deborah McGlauflin, who had been my consultant while organizing the Selling Solar conference. All the people we invited came.

The working papers for the event included a set of case studies SELF had published entitled *Solar Rural Electrification in the Developing World: Dominican Republic, Kenya, Sri Lanka and Zimbabwe*, written by my Kenya-based friend Mark Hankins. This little book had had more impact than it probably deserved on international-development policy makers and governments around the world. In it, we more or less proved that solar was the least-cost option for millions of unelectrified households, and we showed exactly how, in four countries, thousands of families were improving their lives with electric light and increasing their quality of life. Our little report did a lot to demystify the issues regarding electric-power delivery in the two-thirds world.

The October conference opened with a cocktail party on the exquisite garden terrace of Kykuit, former home of J. D., John Jr., and Nelson Rockefeller, overlooking the Hudson Valley. The Rockefellers' Pocantico Hills estate was named a National Historic Landmark in the 1970s. The Brothers Fund leased back eighty-six acres of the estate from the National Trust and converted the barns and stables into an elegant conference center with guest rooms, where most of the attendees stayed. The main meeting hall was furnished with an immense round table surrounded by twenty-six leather chairs. Hanging above it were two twenty-foot-wide tapestries, custom-made for Nelson by Pablo Picasso, the only ones he ever did.

I stayed in the main guest room in Kykuit itself, with views from the second-story windows of the grand entranceway, marble fountain, and huge iron gates, through which most of the world's leaders and its richest industrialists had passed at one time or another. In this august venue, I thought, we had better get serious.

After the opening dinner, hosted by Rockefeller Brothers Fund chairman Colin Campbell (later, president of Colonial Williamsburg), I darkened the room and lit a kerosene lamp to demonstrate how two billion people still illuminated their nights. Then, I doused the lamp and switched on a compact fluorescent lightbulb connected to a battery, the type SELF provided with its solar home-lighting systems. The room was bathed in cool, bright light. In comparison, the glow from the kerosene lamp could not be called "lighting." However, I pointed out that JDR had called his kerosene "the new light" and the "light of the age." Now, I said, *the new light of the age was electric*, powered by the sun.

Century-old cartoon slamming Standard Oil and John D. Rockefeller's global lamp oil (kerosene) cartel.

"We're holding this gathering in an appropriate place," I continued. "We want to bring light to the darkness of much of the third world and replace kerosene with sun-powered electric light. Well, we're right here at the home of a gentleman who became the richest man in America selling *light* to the world in the form of *kerosene*."

When Standard Oil was broken up by a U.S. court in 1912, I explained, over 90 percent of its profits still came from kerosene for lighting, since there was no better alternative. "Now *we're* here at Pocantico to figure out how to give poor people their cheap light. . . . As wonderful as it might have been in its day, kerosene contributes to global warming, creates health problems for its users, and can't power a television or radio. Yet half the rural people in the world will still be relying on a carbon-based lighting source from the nineteenth century as we enter the twenty-first—unless a new direction is taken by governments, development agencies, energy companies, banks, investors, and, of course, the oil companies for whom kerosene production is hugely profitable."

The timing was right for this conference, I said. It came in the midst of many gatherings devoted to the subject of financing decentralized off-grid electrification held in New York, Manila, Morocco, Nice, Sun Valley, and Harare, Zimbabwe (the Solar World Congress). On the other hand, the World Bank's conference Financing Sustainable Development, held the week prior to our Pocantico event, had all but ignored financing sustainable energy. "The majority of World Bank energy economists," I pointed out, "regard three hundred years' worth of coal in the ground as a form of sustainable energy." It has been called "ancient sunlight." The problem with this stored solar energy is that you have to burn it.

I reminded attendees of the recent cancellation of centralized fossil-fuel power projects in Nepal and India—good news—and of controversial projects going ahead in India and China, the bad news. I mentioned how Virginia-based AES Corporation was boasting that the huge coal-fired power plant it was building in Pakistan would bring light to ten million people. I challenged that statement, pointing out that the transmission and distribution (T&D) of the power would be cost prohibitive—and that AES wasn't responsible for T&D. AES's then CEO, Roger Sant, was absolutely convinced that solar energy was too expensive. I would not have believed at the time that ten years hence AES would be one of the world's largest developers of wind power and solar projects.

"I don't think I have to convince too many people in this room that rural people in developing countries will pay for solar electricity in the form of solar home systems," I went on, coming around to what we really needed to discuss at the conference: how to take the money no longer allocated to dams or a thousand-megawatt coal-fired generating plant and use it to finance decentralized renewables, including the most decentralized of all, solarelectric systems for household power. In Nepal, World Bank task managers had decided to focus on hydro and diesel generators . . . completely ignoring solar. This, after the prime minister of Nepal had personally inaugurated SELF's village solar PV project the previous year and had called for a national solar program.

"But then, what's a prime minister against a World Bank energy economist!" I opined.

"If only the world's development economists had listened to E. F. Schumacher when he wrote *Small Is Beautiful: Economics as If People Mattered*," I continued. "I doubt we will think of anything here at Pocantico that he hadn't proposed twenty-three years ago. We need to make 'small is beautiful' big enough to count.

"We're here to look at financing models and brainstorm new approaches and delivery facilities. Not all the rural third world is as poor as you might expect. Many rural farmers in the developing world—living *sustainably*—are better off than many of their compatriots in the megacities. But they have no electricity and they are tired of waiting for it. In the meantime, they are denied all the quality-of-life and transformational benefits electricity offers. They don't want to have to migrate to Kathmandu or Colombo or Delhi so they can see TV, so their kids can read by electric light, so they can quit breathing kerosene fumes. In any case, the lights are going *out* in Kathmandu, Bangalore, Bogota, Manila, and Santo Domingo. Rural people with solar systems often have lights when city people don't!

"We must consider the issue of subsidies, since most electricity and virtually all rural electricity since the first electricity was generated has been subsidized in one way or another—and rural electrification is still subsidized in America. Electricity the world over has always been regarded as a public service."

Finally, I asked, "Will putting fifty-watt solar home systems on one hundred million rooftops help the environment? I think it will.

When solar in the third world is as common as television, the prices will drop and their cities will start using PV, followed by their industries; then, Europeans will use it, and finally even Americans will choose to enter the solar age. And by then huge amounts of carbon will be regularly displaced by clean solar PV." I was reaching here, but a good deal of what I described then has come true. The role third-world PV markets originally played in the development of this industry was immense, even though these early markets have since been eclipsed.

I ended my remarks with a quote from Dr. Hermann Scheer, founder of Eurosolar and promoter of the Global Photovoltaic Action Plan, a sixty-billion-dollar proposal to provide a minimum of ten watts of solar electricity for one billion people:

> Solar energy, the energy of the people, will offer mankind a prospect of survival that would overcome its spreading fatalism about the future. . . . Without a radical shift of the world's energy supply systems to nondestructive solar energy sources, without a solar revolution in the wake of the Industrial Revolution, the *Western model of democracy and capitalism is not the perfection of history but its execution.* (My italics.)

Scheer wrote this in *A Solar Manifesto* (*Sonnen-Strategie* in German), published in 1993.

Scheer, whom you will meet later, died suddenly in 2011. His later books, *The Solar Economy* (2002), *Energy Autonomy* (2007), and *The Energy Imperative* (2012), should be required reading in every university and business school in America.

We discussed all these issues, and everyone contributed his or her expertise. The group concurred that the only way solar energy was going to compete with fossil fuels, or bring power and light to poor people in significant numbers, was if we tapped the great river of capital that, in the mid-1990s, was flowing into the "emerging markets" of the developing world in amounts ten times greater than international donor money. New companies had to be formed to replace the nonprofit charity efforts, like SELF. Philanthropies didn't have enough money to do the job.

The conference was a huge success, for it motivated a dozen people, including me, to try new things. Inspired by this retreat on a beautiful estate amid the great oaks and elms in their fall glory, we all went away on that crisp October Monday morning eighteen years ago trying to figure out how to "sell solar." (The Rockefeller Brothers Fund later published Pocantico Paper No. 2, *Selling Solar: Financing Household Solar Energy in the Developing World*.)

Dr. Jeremy Leggett, an Oxford professor, geophysicist, and expert on climate change (and later the author of *The Carbon War: Dispatches from the End of the Oil Century*), who was then heading up Greenpeace International's solar program, went back to Britain and launched a new commercial enterprise called Solar Century, which was to become one of the United Kingdom's largest providers of residential and commercial solar energy systems. More on Jeremy later in this chapter.

The Grameen Bank executive went back to Bangladesh and started Grameen Shakti, which became one of the most successful solar-electrification companies anywhere, bringing power to over two hundred thousand poor, rural households by 2012.

Richard Hansen, who pioneered solar electrification in the Dominican Republic, went home and began to scale up his for-profit venture, Soluz, which expanded into Honduras.

My friend K. M. Udupa went back to India and helped SELF's new venture there (see chapter 6) develop solar loan programs in association with India's largest banks.

World Bank officials went back to Washington and pressed the bank to finance solar PV whenever and wherever it could.

My neighbor Dr. Peter Varadi, founder of Solarex whom you met earlier, went back to Chevy Chase and launched a global accreditation program (PV GAP) to develop PV system standards so that international lenders would know what they were financing.

The chairman of Solarex went back to Maryland and focused on enabling the company's global dealer networks to better serve and help finance their rural customers.

Dr. Charles Gay, then head of NREL, started a nonprofit company that used solar to enable local entrepreneurs in remote communities around the world sell their products, music, and crafts on the brand-new World Wide Web using solar-powered computers and satellite uplinks. Later, he became a principal in the founding of America's newest solar company, Sunpower.

One "recovering venture capitalist" went back to Washington to found a group to finance small renewable energy companies around the world.

A nonprofit investment bank underwritten by the

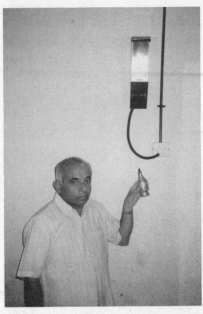

SELCO-India director K. M. Udupa holds traditional oil lamp beneath DC compact-fluorescent lamp fixture. *(Author)*

World Bank, the Rockefeller Foundation, and private funds grew out of the Pocantico meeting. Its founder, who initially brokered environmental grants for the Rockefeller Foundation, had once told me in his New York office, "[SELF] will get checks from the Rockefeller Foundation, but you'll never meet anyone there." Right he was, and yet some very big checks came through for our operations in China and India. One, for a quarter of a million dollars, was lost in the mail, so the foundation just sent me another. As promised, I never met anyone at the Rockefeller Foundation.

John Kuhns, the founder of New World Power, Inc., the first Wall Street–financed renewable energy company (which specialized in wind power and small hydroprojects from China to Brazil to Ireland), went back to Connecticut and thought hard about how to scale up the off-grid PV business in those enticing emerging markets. But John wasn't the type to journey out to remote villages to meet and greet potential customers. He had no real interest in "sustainable business" or "socially responsible investing"; John smelled profits in this vast new energy marketplace and believed noncarbon energy was the future. I would be seeing a great deal more of John, it turned out.

I went back to Washington to figure out how SELF could launch a for-profit solar PV company to undertake what we had now proved could be done, but on a significant commercial scale.

In the spring of 1996 there was a knock on the door of SELF's Dupont Circle brownstone office, and standing there was S. David Freeman, our solar energy guru from the old DOE days, an energy visionary, and an established major player in the U.S. public-utility industry. What the hell was he doing at my door, tipping his rakish cowboy hat?

"Howdy, Neville," he drawled.

I invited him in, and he took a seat in one of the two easy chairs in my office, planting his Tony Lama cowboy boots on the floor.

"I've heard a whole lot about SELF and what you've accomplished," he said. "And I heard you were thinking of starting a company. So am I. I don't think you can do this much longer on a nonprofit basis. It needs to be bigger."

I had been told by colleagues that I should try to contact David, who was then chairman of the New York Power Authority (NYPA), the largest state-owned utility in the nation. As mentioned in chapter 1, President Carter many years before had put him in charge of the Tennessee Valley Authority, the largest United States–owned electric utility. Subsequently, he'd headed up the municipal utilities in Austin, Texas, and Sacramento, California. At the latter he had instituted revolutionary solar energy programs, including the construction of a huge PV power plant that partially replaced energy from a nuclear plant that the good citizens of California, with David's help, had just shut down. In New York he had been trying to institute similar progressive energy conservation and solar programs at NYPA. I knew he was interested in getting NYPA involved with solar energy internationally, and I had once tried to schedule a meeting with him, but it never took place.

Now here he was in my office, uninvited and almost unannounced.

"I thought you were very busy running NYPA," I said

"Governor Pataki fired me. He's a Republican; I'm an outspoken Democrat," David said in his slow, easy Tennessee accent. "Now, I'd like to get back to solar, and maybe I could help you bring solar power to millions of people in developing countries. I'm not too old. I'm sixty-nine. I've got another good ten years left. I understand you've got beachheads established all over the world, but now you need Eisenhower and the troops."

Yes, we had "beachheads," which we called pilot projects. And yes, we needed more help, and now here was America's best-known power-company executive sitting in my office, offering his hand.

This was too good to be true. With David chairing a new company, which I intended to set up and manage, we could raise millions. He knew everyone on God's green earth, from Vice President Al Gore to Maurice Strong, organizer of the Stockholm, Nairobi, and Rio environmental conferences, to Mohamed El-Ashry, chief of the Global Environment Facility at the World Bank, who had once worked for David at the TVA. He knew the CEOs of half the investment banking firms in New York and London and just about every utility chief in the United States.

"Why do you want to do this?" I inquired.

He replied, "I've got such a huge ego that I can't do anything less than save the world." He smiled.

We agreed to look into the possibilities of setting up a company together. I went to New York for meetings with him, and he jumped on the shuttle a dozen times to come back down to Washington to see me and the new lawyer we had hired to help us put a company together. I began examining the requirements for converting a nonprofit organization to a for-profit enterprise and soon discovered it wasn't possible. Instead, I made plans to resign as SELF's executive director and work full-time for the new entity, which we decided to call the Solar Electric Light Company. It was registered as a Delaware corporation in June 1996. It was soon known only by its acronym, SELCO.

SELCO, like SELF, took its name from the Edison Electric Light

Company and the many other "illuminating companies" formed in the United States at the beginning of the twentieth century to generate and sell electric power for the purpose of lighting homes and businesses. Like Rockefeller's Standard Oil, which they soon replaced, the primary business of the Westinghouse and Edison electric light companies was, as their name implied, *lighting*.

S. David Freeman had spent his life working for public utilities: he had never worked for, managed, or started a private company. As a powerful chief executive, he was a man to be taken seriously and was accustomed to having his own way. Beneath the cowboy hat, quick wit, friendly smile, and "aw shucks" demeanor was a take-no-prisoners, steel-hearted manager who commanded total allegiance.

"Neville," he told me one day, "I really don't know much about power or finance or management. I'm really just a psychologist." What he was trying to say was that he read people in thirty seconds and either controlled them entirely from then on, demanding unyielding loyalty, or dismissed them as nobodies. David was quite surprised, despite a six-month happy honeymoon as we brainstormed what our new company would look like, that he could not order me about as he was used to doing with subordinates all his life. I wasn't a subordinate; we were partners, cofounders and co-owners of a private company.

To me, it was simple. David had dozens of other irons in the fire and was busier traveling the country and meeting more people than I could ever imagine doing. Thus, he would be SELCO's nonexecutive chairman, and I would be the CEO.

David had other ideas.

"I'm the boss," he said dryly with a hard edge one day, his disarming drawl failing to charm me this time.

But that wasn't the problem. The problem was that he knew absolutely nothing about the business we were getting into, a business I had already started in India, which was based on six years' experience in eleven developing countries designing and building workable solar-finance models and delivery infrastructures and using suitable technologies. He'd never visited one of these countries to see for himself what it was, exactly, SELF had been doing. On top of that, he had a lot of wrong-headed, in my opinion, ideas, and he rarely followed through on things he promised to do. I guessed he was used to having staff follow up.

What he was good at, however, was *vision*. He had a lot of that. He believed quite sincerely that it was our duty to divert as much Western capital as possible to the developing world to finance SHSs for rural people in much the same way America financed power plants over thirty years by relying on a large customer base of rate-payers. This made sense, to a degree. But we weren't building power plants, and we didn't have ratepayers to finance large capital invest-ments. And we couldn't get thirty-year loans! We only had the potential customer base of millions of poor people whom nobody wanted to finance, at least not over a thirty-year period. Three years was about the maximum loan our customers could expect to get.

"In America, we abolished the hookup charge in most places. You just signed up for power and you got it, provided you paid your monthly bill on time. This is how rural America was electrified. It should be just that simple for all these poor buggers in these coun-tries," he said. I couldn't agree more, except we'd learned the hard way that it was not "just that simple" to transfer David's power-utility model of doing business to the off-grid rural PV markets where we worked. It was impossible to raise the necessary front-end invest-ment predicated on the notion that poor people will pay a "solar utility" to deliver power to them for the rest of their lives. I believed it *was* possible to raise capital on the basis that we would *sell* power systems (and provide service) with sufficient margins to cover the operations, and I believed that we could find third-party lending institutions to make the necessary three-year loans to our custom-ers. In fact, we'd already done that in India.

To investigate how we would raise money, whatever form our business plan took, I arranged to have lunch with David and John Kuhns, whom I had invited to the Pocantico conference. This was a fateful day in ways that would unfold almost immediately and con-tinue to unfold for many years to come.

"I don't think David understands private finance," John said afterward. "He's used to going to Salomon Brothers or Goldman Sachs and getting his millions on a handshake, backed up, of course, by the U.S. government and his base of ratepayers. That's not hard. I don't think he understands the business he wants to get into."

John understood how hard it was to market renewable energy at home and abroad at the time, since his company, New World Power, had already been doing it. Previously, John had formed the largest independent power-generating company in the United States and was operating huge wind farms. He had also financed California's largest geothermal plant.

"If things don't work out with Dave, call me," John said as he left.

Meanwhile, David and I were spending more and more time with our expensive lawyer in downtown D.C., each hour of whose time would buy a poor family in India an entire six-light SHS. Besides crafting corporate documents, the lawyer was noticing, as John had done at lunch, that David and I were not getting along all that well, which did not bode well for a new company. I said, "You're right. This isn't going to work." I began to see that this otherwise well-intentioned man was "all hat and no cattle," as they say in Texas. Vision alone isn't enough.

We parted ways soon after. I got the SELCO name back. David went back to the utility industry, taking a job as chairman and CEO of the Los Angeles Department of Water and Power (LADWP), the largest municipal utility in the United States. He reportedly fired over a thousand people, got LADWP into the black, then turned to his main passion, solar energy. He instituted the biggest utility-driven residential solar PV program in the country at the time, having pushed it through the LA city council and the LADWP bureaucracy.

David moved on to become Governor Gray Davis's "energy czar." He's a hero in California because he kept LADWP—which suffered no blackouts—out of the crooked game with Enron that bankrupted Pacific Gas and Electric and drove California energy prices through the roof. The subsidized solar program he set up was a big success, as he instituted in America exactly the kind of solar business he had advocated for the developing world. This unique American power-house went on to become the manager of the Port of Long Beach and the deputy mayor of Los Angeles. Today, retired at last, he still preaches his solar vision wherever he can.

John Kuhns called when he heard that David and I were not going into business together after all.

"I was impressed with the speech you made at Pocantico," he said.

A well-connected investment banker with twenty-five years' worth of experience in international energy markets, and with an MBA from Harvard Business School, John said he could help me raise money for SELCO. I didn't know how to do this myself, so we joined forces. John had made $30 million by age thirty, lost $60 million, then recouped and made and lost several fortunes since. He was close to my age but came from a very different world that I knew nothing about: the fast-money world of the go-go 1980s and 1990s. I had never made any significant money at all, so dealing with John, a six-foot-two former college football captain who had blown through more money than I could ever make in five lifetimes was, indeed, intimidating. I did learn, however, that people make money who want to make money; I had never cared about money, so I didn't have much. But soon I was to have a company and was rich on paper!

I named Bob Freling the executive director of SELF and resigned from the organization to devote myself full-time to SELCO. The company acquired from SELF the in-country infrastructure in four countries on which it could base its new commercial activities. It would acquire SELCO-India, which would become a subsidiary for which we could now provide investment capital directly.

John and I worked up an ambitious business plan that we flogged to the big investment houses in New York, Dallas, and Boston. John got the doors opened, but soon they were shutting rapidly behind us as Bear Stearns, AIG, the Bass Brothers, and other billion-dollar "emerging market" funds politely laughed us back into the street.

"Solar? Hmmm. Third world? You've got to be kidding!"

We tried to explain in our presentations that these countries offered the biggest untapped power market in the world, and SELCO was going to tap it with solar technology. Solar, we explained patiently, was the only thing that would work where the power grid had failed or did not yet exist. This was big!

I watched a lot of eyes glaze over as I pulled out photos of smiling poor people standing in front of their solar-powered houses. "We're not going to invest in a charity!" one annoyed fund manager said. We explained that although a nonprofit organization had teed up this business opportunity during many years of pioneering development work, SELCO was a totally independent, for-profit commercial enterprise that expected to make a lot of money for its investors . . . one day.

No takers.

Meanwhile, I called Jeremy Leggett in London, the Greenpeace rep at Pocantico, who had become a friend. He invited me to attend the Oxford Solar Investment Summit, which he was organizing. Dozens of the world's largest insurance and reinsurance companies would convene to figure out how to invest in solar energy instead of the carbon-fuel industry, since they had by now all experienced the results of catastrophic climate change attributed to carbon emissions. During the 1990s, the biggest weather-related disasters in history had taken place: Hurricanes Andrew and Mitch, freak storms in Europe, killer heat waves in Chicago and India, and floods in China and Bangladesh. Over the twentieth century, a billion-dollar insurance disaster usually happened only once a decade; now, to the insurers' and reinsurers' despair, billion-dollar weather disasters were occurring every year. Our concerns were valid and prescient: in 2011 the United States alone was hit by *fourteen* weather-related disasters that cost $1 billion or more each. Katrina had cost $108 billion dollars. Hurricane Sandy wrought $71 billion worth of destruction in eight states. In 2012, Swiss Re cited U.S. weather disasters for $77 billion in insurance losses.

The Oxford Investment Summit in 1996 brought together over one hundred U.K. and European insurance CEOs, bankers, pension-fund managers, and solar company executives. It was coorganized by Rolf Gerling, chairman of Germany's insurance giant, the Gerling Group. Rolf Gerling was then listed by *Forbes* as one of the top three hundred billionaires in the world. Shortly after the conference, Gerling's Swiss publishing company produced a book, edited by Jeremy, entitled *Climate Change and the Financial Sector: The Emerging Threat—the Solar Solution.* The book was an outgrowth of a 1995 gathering of business leaders in Berlin on the eve of the Climate Summit, where one hundred fifty nations gathered to negotiate the Framework Convention on Climate Change and launch the Intergovernmental Panel on Climate Change (IPCC), which eventually resulted in the Kyoto Protocol (later followed by the disastrous Copenhagen meeting in 2011 and the even worse Rio + 20 conference in 2012).

"We were here in Berlin," Gerling wrote, "because most of us were worried about global warming and the economic disruption that may accompany it."

Climate Change and the Financial Sector outlined, from the perspective of leaders of the global financial sector, what would happen if global energy demand were met exclusively with fossil fuels. Since insurance and reinsurance companies were the largest single investors in the carbon fuel chain, especially in fossil-fueled power plants, it was paramount that they change their thinking and find new avenues of investment that would help reduce greenhouse gases, which were blamed for the climate disasters threatening the world . . . and the insurance business. As far as these insurance and business leaders were concerned—this was in 1997!—no more studies were needed on global warming and climate change; they already knew what was happening. Now they proposed that their industry be part of the solution instead of part of the problem.

Jeremy Leggett addressed the Oxford gathering, saying, "The core of a clean-energy future requires a huge component of solar energy. Massive solar markets offer the key to a sustainable energy future. . . . A solar revolution begins to address the population problems, the clean water problem, air quality, urban migration, and nuclear proliferation, among other societal threats."

I delighted in hearing these words. It was "solar revolution" time once again, words I had not heard since our heady days in the Carter administration seventeen years earlier. Now, would any of these financial types be interested in investing in the Solar Electric Light Company?

Yes, Rolf Gerling would. Thanks to Jeremy's introduction, Gerling agreed to look at our business plan.

Jeremy also opened doors with another important billionaire. Stephan Schmidheiny, of the famous Swiss industrialist family, was a behind-the-scenes global activist who had financed NGO participation at the Earth Summit in Rio in 1992 and had also founded the World Business Council for Sustainable Development. Schmidheiny's business manager agreed to look at our plan.

Jeremy's business task force on climate change in Berlin had concluded that PV would be very important in the future, *largely because of the markets of the developing world.* He had written, "A key global imperative . . . will be the delivery of sustainable energy for development in the developing world, where two billion people currently have no electricity; the most realistic form of alternative supply, especially in rural settings away from current electricity grids, is solar PV."

This argument won over Switzerland's richest man and biggest environmental and international-development philanthropist. SELCO was there, waiting to get into these off-grid markets in the developing world. We were not looking for philanthropy but straight commercial investment made strictly on the merits of our business proposition.

In the summer of 1997, John Kuhns and I flew to Europe for meetings in Cologne with Gerling's people at his new investment fund, GAIA Kapital, and in Zurich with Schmidheiny's representative, as well as with a Swiss public investment fund that was raising "green money for a blue planet."

By August SELCO had successfully raised its first $2.5 million of working capital, and we were ready to go. We had struck out with venture capitalists and private equity, so that only left angels and institutional investors. Thanks to Jeremy, I found the latter, but I never could have closed the deal without John Kuhns, whose knowledge of "financial engineering" was dazzling. He believed in our business, but he didn't understand how it worked. I knew how it worked, but didn't know how to make money at it. He'd always hoped for an IPO (initial public offering), since that's how he'd raised money for his previous successes at Catalyst Energy and New World Power as well as his bond trading firm in New York. But IPOs were dead after the dot-com bust in 2001. John, long after leaving SELCO's board, invested in hydroelectric companies in China.

We opened a new office in Chevy Chase, Maryland, that I could walk to from home. I didn't want a local car commute on top of my regular ten-thousand-mile transits to India, Sri Lanka, China, and Vietnam.

I named John Kuhns, Paul Maycock, and another Harvard MBA and Wall Street financier to my board of directors. I hoped these board appointments would reassure our billionaire investors that we were serious about this business. I also felt that I needed help because I had no experience managing a company on this scale. However, my investment bankers had never managed a small start-up like this, and they would never really come to understand SELCO's unique mission. We were going to break new ground, learn by doing, and come up with our own formula for creating a profitable, high-growth, but sustainable business that would be a nonextractive and nonexploitive business model for the developing world.

I discovered that the world of Wall Street finance is about as far from the ideals and practices of SELCO and its belief in sustainable social service as you can get. I was to discover that most MBAs didn't see the world in the same way as our risk-taking billionaire environmentalist investors or our mission-driven operations managers in the field, who were responsible for actually running the company and generating revenues. Henry Mintzberg, a McGill University professor, argued at the time in his book *Managers Not MBAs,* "All MBA graduates should have skulls and crossbones stamped on their foreheads, along with warnings that they are not fit to manage." He criticized the MBA view of the world as "calculating and analytical" and totally lacking in any understanding of management. "The smartest guys in the room," the guys who brought about Enron's collapse, were mostly Harvard MBAs. So was Enron's good friend George W. Bush, who somehow got a Harvard business degree, making me wonder, just how difficult could it be? An MBA didn't keep him from running our country into the ground.

In the next six years, SELCO walked a tightrope as it sought to become a "new paradigm" minimultinational corporation beholden to the triple bottom line of "people, planet, and profits," more clearly stated as social *responsibility,* environmental *sustainability,* and *profitability.* It would take all our *abilities* to achieve this, and I would begin to question whether ethical business practices, socially responsible capitalism, and the principles of sustainability can ever be reconciled with the demands of finance capitalism.

What we didn't know at the time was that we had pioneered what came to be called in the twenty-first century "social enterprise." We had no name for what we were trying to do and subscribed to the "just do it" approach. We had a lot to learn, and with our first couple million of capital in the bank, we were on a roll. But idealism can't always win over established business practices, and today I'm amazed SELCO, now based in India, is alive and well (see chapter 6), since I confess to making just about every mistake in the book along the way. Learning by doing.

Meanwhile, I was to discover that I loved business, free enterprise, fair trade, and working in a market economy but that I didn't like Wall Street bankers, international finance, so-called free trade, and what modern capitalism and its manipulated and managed markets had become. I had read Paul Hawken's 1993 *Ecology of Commerce,* a landmark work and a turning point for me. Hawken's book was the original inspiration for my deciding to leave the genteel world of NGOs and join the demanding and challenging world of business. (Years later, to my surprise, Paul would call me up to seek my support and advice for a brand-new solar technology venture he had launched.)

I would soon experience firsthand the disparities and contradictions inherent in the global growth economy. I had not yet read David Korten's 1999 book, *The Post-Corporate World,* in which he states, "The global economy is being centrally planned for the primary benefit of the wealthiest one percent of the world's people, a triumph of privatized central planning over free markets and democracy." He described America's "one percent" twelve years before the Occupy movement erupted in the United States, representing the

"99 percent." SELCO wanted to represent "the people," not the global elite; our values were with the young people from the developing world, with whom we were organizing a twenty-first-century ecoenterprise that would not only bring light to hundreds of thousands of people but would itself be a beacon for anyone seeking a new way for market capitalism, free enterprise, and fair trade to serve the world— and maybe even *save* the world. Bringing power to the people would prove to be more fun than I deserved to have.

A year after I resigned from SELF, leaving it in the hands of Bob Freling, the organization was nominated to receive the Millennium Award for International Environmental Leadership from Global Green USA and Green Cross International, chaired by the former president of the Soviet Union, Mikhail Gorbachev. "SELF's commitment to providing a self-sufficient renewable energy path for developing countries is remarkable," read Global Green's announcement letter. "The spirit of the Millennium Awards is to inspire others around the world to follow SELF's cue."

Among the other recipients were CNN founder Ted Turner and his environmental foundation and Sir John Browne, chairman of BP and a strong proponent of solar. Bob flew to LA, where Mr. Gorbachev presented the award to him at a gala event at the Four Seasons Hotel. Bob was able to thank Mr. Gorbachev in fluent Russian. My wife, Patricia Forkan, also attended this glittering event, as she still served on SELF's board. Patti shared a table with Paul Maycock (SELF's new chairman) and SELF's newest director, *Dallas* star Larry Hagman.

I was in India, tramping around the rural areas of Karnataka and organizing SELCO-India in Bangalore, so I missed it.

6

Sacred Sun at Your Service: 1993–2002

On a very cold day in January 1993, Harish Hande climbed the three flights of stairs to SELF's office on Connecticut Avenue and shyly entered, took a seat, and said nothing. He was wearing only a thin shirt and no undershirt, and he did not even have a jacket with him.

"You must be freezing," I said.

"I'm okay," he said, dismissing my concern and introducing himself.

Maybe he's some sort of yogi who doesn't feel the cold, I thought. After the brief introduction he continued to say nothing. He was very serious and had an aura of great intention and purpose about him, but I wasn't sure exactly what it was.

Harish called me Mr. Williams, and I mistakenly corrected him, preferring the informal American style of false, friendly familiarity that most foreigners find strange. He absolutely could not call me by my first name, not someone nearly the age of his father, as such familiarity would be considered extremely disrespectful to elders. I never regarded generation gaps and age differences as mattering very much, but of course they do. Since I forbade Mr. Williams, he never called me anything, and thus our business relationship was born. He could only stammer out what he was trying to say without prefacing it with Neville or Mr. Williams. At least he never called me sir.

However, his awkward shyness didn't stop him, for he wanted very much to work with SELF and find a way to bring solar rural electrification to his home country, India. And he had fire in his eyes.

Harish (I'll call *him* by his first name) was studying for his doctorate at the University of Massachusetts, Lowell, under the only professor in the United States offering for-credit courses in solar electricity and solar engineering. He was a graduate of the prestigious Indian Institute of Technology (IIT) at Kharagpur, the Indian university group that is harder to get into than Harvard or Yale. He had recently trained with my friend Richard Hansen in the Dominican Republic, where he had seen "the light"—solar light. He wanted to take Richard's model to India, but Enersol didn't have any money or interest in India, so he'd come to see me.

I said that the people at the World Bank's Asia Alternative Energy Unit had been after me to think about starting a solar project in India, and that the UNDP's India representative had begged me to come to New Delhi when he heard about our success in Sri Lanka.

In any case, I had no immediate plans to enter India with a solar project because the size of the place intimidated me, and I wasn't sure how or where to start. On my first trip there only two years earlier—on a brief sightseeing excursion from Sri Lanka—I had been driven right by the spot where, a week earlier, Prime Minister Rajiv Gandhi had been blown to pieces by a female suicide bomber from Sri Lanka's Tamil Tigers. The blood-covered folding chairs were still strewn about the horrific crime scene, which was not even cordoned off. Overlooking the public park where the bombing had occurred was a large statue of Indira Gandhi, Rajiv's mother, who had also died violently, gunned down by a Sikh bodyguard. Her stone visage had gazed down from its tall pedestal on her own son's murder. As a result, I was beginning to think the third world was a pretty dangerous place and wondered if India would be stable in the aftermath of Rajiv's assassination.

Harish said not to worry. India was stable. It was a democracy.

I sent Harish Hande to India during a school break to find a good location to begin a pilot solar-electrification program for SELF. For one month he traveled from north to south and east to west. He had lived in Gujarat and Orissa and had family in Karnataka. His father had been the manager of India's government-owned steel plant, Essar, and the family had moved around a lot. Harish spoke six

Indian languages. He and his family were Brahmins, of course, the priestly caste in India.

Harish came back to Washington to see me that spring and announced we could do a project in Karnataka, formerly a "princely state" of the maharaja of Mysore, renamed in 1973. It comprises the north end of the Malabar Coast's warm coastal plains, the cool highlands amid the Western Ghats, and the southern part of the Deccan Plateau, four thousand feet above sea level, which is the site of Bangalore, a midsized Indian city of (then) eight million people.

"Why Karnataka?" I inquired.

"Because it is India's richest state and has the worst power cuts, and because I have family in Bangalore and Mangalore," he said matter-of-factly. (Mangalore is a "small" Indian port city of one million inhabitants.)

His assessment was good, but he didn't have any idea how we'd get a project started there. SELF always required local partners, community organizations, and lenders. He didn't know any, nor did I.

That didn't stop us. I went to India shortly thereafter and joined Harish in Mangalore. He introduced me to his remarkable aunt, Hemalata Rao, a self-taught entrepreneur who ran a small cable-TV company out of a shed behind her house. The biggest barrier to her expansion was not a lack of customers, but a lack of customers with electric power. Her cables reached the fringes of the rural areas, where power was intermittent. She understood the need to bring power to the people, which she assured me the perpetually bankrupt State Electricity Board would never do. Ironically, I arrived during Diwali, the Hindu Festival of Lights, which the city managed to keep lit most evenings.

Mrs. Rao took me into her "TV studio," where her stacks of amplifiers, transmitters, and VCRs were the source of a half dozen channels, including her own movie channel. Overhead, in the rafters, a fourteen-foot-long black whip snake slithered and hovered, looking down at me. "Oh, he lives here," she said. "He eats the rats." Mrs. Rao was a follower of Paramahansa Yogananda, the Indian guru, or teacher, who, following Vivekananda, came to America in the 1880s and brought Hinduism to the United States in the early part of the twentieth century, settling eventually in southern California. I had read Yogananda's perennial classic *Autobiography of a Yogi* some years before, and his spiritual journey had made an impact. Portraits of Yogananda and other Indian saints graced her workshop. She had named her company Anand Electronics, incorporating this widely used spiritual appellation.

"How big a territory are you licensed to cover?" I asked her.

"We don't have licenses. I just extend my cable service as far as I can reach until I meet the next cable operator's customers," she said. "Whoever gets there first gets the customer. That's how it works."

India, I learned, had some sixty thousand independent cable companies like hers in the early nineties. For a formerly closed socialist state with a suffocating bureaucracy, India was inherently a pretty entrepreneurial place. The market economy at this level had not been harmed by Jawaharlal Nehru and Indira Gandhi's (Nehru's daughter) move to nationalize heavy industry, utilities, banks, and airlines. In fact, it thrived. It occurred to me right then that if we could harness this grassroots, free-enterprise, business-minded culture to provide solar electricity, we could be very successful.

The Indian electricity problem was bigger than anyone knew or wanted to know. It was especially severe in Karnataka. "There is no more expensive power than no power," Mrs. Gandhi had said when she sought to continue her father's work of electrifying all of India in the spirit of Roosevelt and Lenin, who both put subsidized electrification at the top of their national agendas. But India had no oil and no economic way of transporting coal, and the government's stultifying bureaucracy and socialist policies kept new generating plants from being built.

India was a well-wired country with no "current," as they referred to electricity, lots of powerless power lines, and a grid infrastructure that might bring one light to the police station in a small community, which would then be marked on a map in New Delhi as an "electrified village." I was amused by a cartoon in the *Indian Express* that showed a fat, traditionally garbed Indian politician visiting a community and exclaiming, "What! This village has no electricity? But I promised it myself ten years ago!"

So in this, the world's largest democracy, the people would elect new politicians, who would also promise electrification, and again it would not come. Life went on in this great land, and government officials boasted that 90 percent of Indian villages were electrified, while one hundred forty million families, in actual fact, did not have electric power. Government-subsidized, mostly imported kerosene continued to be the lighting method of choice. Harish and I certainly had our work cut out for us here.

We flew from Bangalore to the twin commercial centers of Hubli-Dharwad on the fertile, cool Deccan Plateau, where the British viceroys had originally planned to put the capital. These were large, dusty, backwater towns of half a million each, serving the agricultural regions surrounding them. In Dharwad, Harish introduced me to K. M. Udupa, a family friend who had taken a liking to Harish and what he was intending to do "for the people of India." Mr. Udupa was chairman of Malaprabha Grameena Bank, a regional rural-development bank with 210 branches, which was owned by Syndicate Bank, one of India's largest government-owned commercial banks. Malaprabha Grameena Bank served over 1,100 villages. Mr. Udupa was in the process of building a modern headquarters with solar-powered backup systems and solar-powered fountains and streetlights. Harish had been hired to install the solar modules.

Over dinner at Mr. Udupa's—I dared not call him by his first name, and never have since, since he was three years older than I!—we talked about our respective backgrounds. His wife and daughter hovered in the kitchen, periodically bringing out dishes of savory food. Women never ate with the men.

"I visited the United States once," he told me. "It was 1971. The trip was sponsored by the State Department as part of an agricultural exchange. As a rural banker, I was interested in animal husbandry."

I asked him where he went.

"Ohio," he said.

"I'm from Ohio," I replied. "Where in Ohio?"

"Oh, a small place. You probably never heard of it."

"Try me," I said.

"Chardon, a small village east of Cleveland. Near the Amish communities. We studied their farms."

I smiled and said, "That's where I'm from! I went to Chardon High School. I worked with Amish carpenters in the summertime, had to go to their farms to pick them up since they don't drive."

"Yes, they still use horse and buggies," he said, amazed at this being the case in modern America.

He proceeded to bring out his scrapbook, which contained a photo of Mr. Udupa and my old high school principal, who was on the Rotary Club's committee that had welcomed the Indian delegation. I don't need to mention the odds against this happening. We laughed and became fast friends. Mr. Udupa later joined the board of our Indian company.

K. M. Udupa, a short man always nattily dressed in a short-sleeved polyester safari suit, was Indian's pioneer of biogas digesters. For the uninitiated, these are large, well-like, concrete enclosures that process cow manure and turn it into methane gas that can be burned for cooking and lighting. Tens of millions of them had been sold in India, and millions had been financed by Syndicate Bank and its rural subsidiaries. They were the ultimate example of "sustainable energy"—provided you had cows.

"We need to do with solar power what we did with gobar gas," he said, using the Hindi word for cow manure. "There is no other solution to India's rural power situation except solar, and Harish Hande here, I believe, is going to make it happen, with your help."

This was a tall order, but I accepted the challenge of helping Harish fulfill his dream, and Mr. Udupa's dream, of providing solar lighting to every rural family that could afford it. No giveaways or subsidies— this would have to be purely commercial, we knew that. Just like gobar gas.

"You know, in India, the sun is considered sacred in Hindu spiritual life. It's name is 'surya,' from the Sanskrit. We have worshipped the sacred sun for thousands of years," Mr. Udupa informed me. "We understand the sun's importance in our lives. People already understand solar energy, naturally."

The next time I met Mr. Udupa was in his office at his new posting in Hyderabad, as general manager of Syndicate Bank's main city branch, a big promotion from rural Dharwad. Here, he told me the story of India's fifth-largest bank and how it started as a cooperative venture in 1925 in Udupi, a small village near Manipal, Karnataka, a university town a half hour north of Mangalore, which was where Mr. Udupa was born and grew up. Whereas most large national banks were headquartered in New Delhi, Bombay (now Mumbai), Calcutta (Chennai), or Bangalore, Syndicate Bank's head office was in tiny Manipal, on a hill overlooking the Indian Ocean. This is the bank that invented "microcredit" to serve rural people. Mahatma Gandhi's portrait hangs over the desk of every Syndicate Bank manager.

The bank's founder, the Pai family of Manipal, had a close relationship with Prime Minister Indira Gandhi, who admired the family's enterprising ways and visited Manipal often. But this did not prevent her from nationalizing the bank in 1975 during "the emergency" in India, when she suspended civil liberties and ruled by decree for nineteen months in response to political opposition. Nonetheless, the Pais continued their development crusade, founding large engineering and medical colleges in Manipal. When I was introduced to the Pai family in 1995, all they wanted to talk about was bringing solar electricity and hot water to the college dormitories, where nightly power cuts prevented students from studying. We later electrified many campus buildings in Manipal while installing central solar hot-water systems atop the dorms.

Syndicate Bank managers wanted to electrify their rural bank branches, where power for lighting was rare or nonexistent. We started with the money-counting rooms and vaults, and the bank paid cash for the PV systems. SELF wasn't too interested in electrifying banks; we wanted to electrify households. However, the banks were the key.

I was trotted around to speak at many solar-lighting seminars in South India that were sponsored by Syndicate Bank and attended by hundreds of rural branch managers and local government officials. It was usually a big deal to have an American there, and they always made a fuss. I felt a little like a British colonialist, except that I was learning more from these people than they could ever have learned from me. Mr. Udupa introduced me to several powerful chairmen of South India's largest banks; I would be ceremoniously ushered into the great man's enormous chambers, while peons, aides, and numerous supplicants hovered outside, seeking an audience. Mr. Udupa, Harish, and I would pitch our mission to bring light to rural people, and the chairman always listened carefully before launching into a lecture on rural development and what to do about it.

Mr. Udupa would say, "Ninety percent of the households in India do not have electricity, but the government says ninety percent of villages are electrified." The various chairmen and the ever-present obsequious managers would then converse with Mr. Udupa in rapid-fire Indian English or the local language, and my presence would be forgotten. Having made my opening remarks, I gazed out the windows or studied the colorful wall calendars with their illustrations of India's pantheon of gods, took tea when offered, and otherwise remained mute. Afterward, Mr. Udupa would always say of the chairman, "He's with us."

We launched a multistate training program through Syndicate Bank, SELF, and USAID, which later provided funding support. We educated thousands of bank managers over the years to understand what solar electricity was, how our small household PV systems (SHSs) worked, and how they could be securely financed at the village level.

The multibank program was later highlighted in a gilded ballroom at the posh Oberoi Hotel in Delhi during a United Nations workshop on "credit mechanisms for renewable-energy applications" in 1997. All the bureaucrats gave their monotonous speeches about how much the government and aid agencies were doing to support solar energy (precious little, despite India having a government agency and ministry dedicated to renewable energy). Then, out of the blue, Harish arrived with the chairman of Syndicate Bank, who gave his *own* speech about making solar loans at the village level. The chairman arrived in a large white Mercedes, flags flying on the fenders, with two motorcycle-police outriders. He upstaged the UN's "alternative energy" officials at their own conference by describing how his bank's rural network was actually financing solar power for poor people and not just talking about it.

Harish and I retreated to my hotel room in Bangalore in October 1994 to figure out how SELF could launch a solar project in India. We pondered and talked and kicked around ideas. Having now seen much more of India, I was no less intimidated by the challenge, despite having met so many amazing, entrepreneurial, and visionary people with the interests of rural India at heart. Whom could we trust as local partners or institutions to bring solar electricity to rural people? We had not yet found an NGO to partner with, and we didn't want to set up our own nonprofit group, for that wasn't SELF's approach.

Then it hit me: we could "commercialize" this endeavor right from the start in South India by drawing on the local entrepreneurial spirit and talent and *simply forming a company*. Harish liked the idea immediately. Now what?

I knew India was going to be a difficult place to do whatever we decided to do. The country had been economically and financially closed to the world since independence in 1947, becoming self-sufficient, if stultifyingly inefficient, in almost all things. Foreign exchange laws were the most restrictive on earth; you couldn't send money in, you couldn't take money out, and the rupee had never been freely traded. But this was the 1990s, and globalization and the global economy were beckoning, for better or worse.

Rajiv Gandhi had sought to liberalize India and open it to the world economy while promoting and encouraging foreign investment. After his assassination, it was not immediately clear what his successors in the Congress Party would do. But the next prime minister, Narasimha Rao, took the bold step of keeping the country on the path to economic change and liberalization. This meant foreigners could own 51 percent of an Indian company (but not power companies or restricted industries). Foreign companies were not allowed to do business directly or to open branch offices; instead, they had to either form a joint venture with an Indian company (as BP Solar did with the Tata family in 1989, forming Tata BP Solar, once the largest manufacturer of solar cells and modules in the developing world) or set up a new entity, of which they could own a simple majority of shares.

Harish and I took a three-wheel Bajaj taxi across town to meet with attorneys at one of India's oldest law firms, King & Partridge, something of a holdover from the English barristers who had bequeathed the country a strong foundation in English law.

I have no gift for languages and can barely understand French, or speak it, despite long study, a year at a French university in Switzerland, and a stint driving a taxi in French-speaking Montreal. But I seem to be able to understand accented English. There are hundreds of versions of English spoken worldwide, but no one speaks faster and with more idioms and odd syntax than Indians, as Americans would learn when the call centers came into being. Anglo-Indians and the upper classes, of course, speak pure Oxfordian English with round vowels that put some BBC announcers to shame. But most Indians speak another English dialect entirely (besides their natives tongues of Hindi, Urdu, Kannada, or Malayalam). Communication was everything, and if my tin ear for foreign languages had applied to foreign English, I might as well have packed my bags and gone home. Moreover, Indians often could not understand me except when I spoke very, very slowly, in crisp clear tones. American accents put them off.

Anyway, my understanding of colloquial Indian English came in handy at the law firm as the intense "advocate" fired off the steps required for a foreign entity to acquire or form an Indian firm. King & Partridge agreed to incorporate our new company in India and register us as a foreign-owned firm, requiring special permission from the "permit raj," as India's layered bureaucracy is called.

The Solar Electric Light Fund was a nonprofit and, as such, couldn't be incorporated under the new investment law or any other laws of India, so we needed to form a new commercial enterprise that SELF would manage, but it had to be separate and independent.

We needed, first, to come up with a name.

As mentioned in the previous chapter, we came up with Solar

Electric Light Company because electric light was our main product. Solar Electric Light Company, as a spin-off from SELF, sounded about right.

Names usually get reduced to acronyms, so we became SELCO, doing business as SELCO-India. At last, on March 14, 1995, King & Partridge got India's first foreign-owned solar energy company approved by the Indian government. I went back to Washington and began raising money through our new American holding company, SELCO-USA, as discussed in the previous chapter. The first investments came from the aforementioned private nonprofit investment fund sponsored by the World Bank, the Rockefeller Foundation, and the Rockefeller Brothers Fund. When Indian laws changed in 1998, the American parent company was able to own 100 percent of SELCO-India, which it managed to do after a lengthy appeal to the president's Foreign Investment Promotion Board in New Delhi. It would be the first 100 percent foreign-owned solar company in India.

Harish, now finished with his course work in Lowell, moved back to India in 1994 and got down to business while he completed his doctoral dissertation. He traveled all over South India setting up the first SELCO solar service centers, the village-based outlets that would become the basis of the business. He trained local youth and hired technicians, managers, and salespeople. He worked tirelessly to establish our new enterprise. His aunt, meanwhile, began producing charge controllers and DC fluorescent lighting fixtures in her workshop. Harish was inspired by one woman he met, who bowed down and touched his feet and told him she wanted to see electricity in her house before she died. "What was remarkable," he told me, "was that she would pay for it!"

On my next visit to India, Harish introduced me to an illiterate, self-taught TV repairman from a small town near Mangalore, who quickly took on the mantle of the SELCO mission. If I approved, Harish would hire him as the company's first manager. Here was an Indian Brahmin and graduate of India's most prestigious college sitting down with an unschooled village appliance repairman to build

SELCO-manufactured DC "luminaires" provide a variety of bright indoor and outdoor lighting; LEDs have subsequently been added to the mix of options. *(SELCO-India)*

our little company. How could I say no? Almost entirely on his own, Umesh, as he was called, created SELCO's first sales and service center, personally made the first several hundred sales—without a phone or his own transport—and oversaw the installation of each and every SHS in a twenty-mile radius. Umesh was a consummate salesman, believing in what he sold, and local people trusted him. These are the kind of human beings that development professionals and foreign aid consultants never meet.

On one of my visits to India years later, Umesh, in his limited English, tried to explain to me how he kept track of exactly how many liters of kerosene we displaced and how he calculated the resulting reduction of CO_2 emissions from kerosene lamps. He showed me our total tonnage of "carbon offsets," and we were later able to sell existing and future offsets for a half million dollars, becoming the

Forty-watt solar panel
electrifies farmer's home
in Karnataka, India.
(SELCO-India)

first solar company in the world to find a market for carbon offsets from solar power, experimental though it was.

SELCO-India didn't have money yet for its own transport, so Harish rented jeeps and trucks as necessary, and all employees used their own motorbikes. One day we rode a jeep through some wild hilly country to visit the company's very first SHS installation, in the house of a local landlord who had twenty-five peasant families working for him growing coconuts and cashews. I snapped a photo of his wife turning on her new living room light. She was so excited. She said, "You must send a photo to my sister in Arlington, Virginia."

The best thing about that day was seeing that the system worked using our own electronics and lights, powered with a PV module made in India. It was working a decade later, with only one change of battery.

Our first family of solar was so happy with their system that they wanted all their sharecroppers to have solar lights as well, so we established the first revolving solar loan fund in India with a grant from SELF. Soon after, room and porch lights could be seen at night glowing from twenty-five peasant homes scattered up and down the verdant hills. The loan fund's operation proved to Mr. Udupa and to local Syndicate Bank managers that people would pay for these systems on installment credit. Our salesmen began signing up more customers and taking them to the rural bank branches to help them negotiate three-year loans for $275, the amount after the down payment and front-end cost of wiring the house. This was close to the average annual per capita income of Indians then, so for them buying a sun-powered electric system was like our buying a car.

Five years later, one Indian banker told me, "solar lighting systems are just like any other consumer item, except that they have social benefits and qualify for antipoverty, 'priority-sector,' lower-cost loans. We treat solar as a durable good, like a bicycle or tractor or refrigerator or television."

We were now where the American auto industry was when banks first made loans on the horseless carriage, confident that the device would have a useful life at least as long as the life of the loan. We promised the banks in writing that we would buy back any systems they had to repossess, but this rarely happened.

Bangalore was to be the site of the company's head office. It is the capital of Karnataka, a state of 43 million people! It is also India's nicest city, with tree-lined boulevards, vast public parks, grand government buildings, and a pleasant climate year-round. But the city was the victim of daily power cuts, scheduled and unscheduled blackouts, that were beyond the ability of the hopelessly inefficient Karnataka Electricity Board to avoid. Bangalore was already the Silicon Valley of India (and would later become the capital of American-owned call centers). How was this possible without reliable electric power?

It was possible because every business (and every luxury hotel where foreigners stay) had—and many still have—their own diesel generator, thus exacerbating the pollution problem. Nothing could be more hopelessly inefficient and costly. Meanwhile, rural people, even urban people, remained in the dark. I met the chairman of the Karnataka Electricity Board and sought to interest him in solar, but he had other fish to fry, such as keeping his aging generating plants going and his transmission lines maintained. Solar energy was a dream to be deferred, thank you. Anyway, he informed me, 85 percent of Karnataka's village was already electrified, so not to worry, they'd soon get to the rest. Indian bureaucrats are victims of their own misinformation, and the higher the official post, the bigger the prevarication. In fact, 50 percent of Karnataka's peasant farmers had no household electric connection, whether they lived in villages or on widely dispersed farms, even if they could see distribution lines along the nearest roadway. Thus, SELCO-India advertised "electricity at your doorstep."

This was the era of "emerging markets" in the go-go 1990s, when every power utility in the West wanted to build generating plants in India. They spent millions trying to navigate the Indian bureaucracy to secure permits. Huge American firms sent in their naive minions, and signs went up announcing where new power plants would soon arise on expensively acquired or leased vacant land.

Mr. Prabhakara, the general secretary of Karnataka (the highest civil service office, while the governor holds the highest political office), told me that the state's electric crisis would soon be eased, and he itemized the new megawatts of power generation that would be coming online thanks to foreign investors. But they never did. When he retired from the Indian civil service, a disillusioned Mr. Prabhakara became one of SELCO's biggest supporters and worked closely with Harish to extend our "solar services" to more rural communities.

We were all watching Enron's plans in Maharashtra, the neighboring state to the north, where Enron and Bechtel were building a 2,000-megawatt gas-fired power plant called Dabhol. Rebecca Mark, the aggressive Texas executive who was Enron's project manager, was then a celebrity in India and appeared on the cover of newspapers and magazines. When Enron's corruption and fraud brought one of America's wealthiest public companies to its final collapse, U.S. taxpayers lost $700 million, the amount the Overseas Private Investment Corporation loaned the Dabhol project, while Enron investors lost everything. Ms. Mark, described by *Fortune* magazine as "one of the 50 most powerful women in American business," cashed out of crumbling Enron in 2000, sold her stock, and pocketed $83 million, a pretty nice reward for failing to bring electricity to a single Indian household. Enron's corporate honchos could not understand that no Indian government could raise the electric rates high enough to finance new power plants with foreign investment. Furthermore, Enron wasn't interested in distributing the power, only generating it.

We always knew we'd outlast Enron, especially after I heard about Enron's plan to build a 50-megawatt solar photovoltaic power plant in Rajasthan. (Enron had purchased half of Peter Varadi's Solarex.) Who was going to distribute all this power if this central solar plant were built? Certainly not Enron. This was an idea too early and unworkable for its time, although large central, utility-scale PV plants would eventually be built in India.

The best thing about solar energy is that it is already distributed and can be generated on-site, where it's needed. *This basic fact escapes most energy planners and power engineers.*

But back to our story.

Before Harish could afford an office or staff in Bangalore, we needed additional capital. About that time my phone rang in Washington. It was USAID, offering me $400,000 for our project in India.

I proceeded to draft a budget and write up a proposal. After lots of haggling, after several consultants took their cut and USAID contractors took nearly half the money off the top for "project administration," SELCO-India received a check in the mail for U.S.$160,000. Harish was incredulous when he opened the envelope. He never really expected one dollar. This would buy a lot of rupees. The only problem was that the USAID funds were in the form of a "conditional grant," which was actually a loan that SELCO would have to pay back. We did.

The conditional grant funds allowed Harish to open SELCO's first office in Bangalore and to hire two experienced "solar salesmen" and business managers who had been involved with solar for several years. They designed a big sign for the suburban two-story headquarters. Harish continued his peregrinations to the far reaches of Karnataka, Andhra Pradesh, and Tamil Nadu, where he indefatigably promoted "sun at your service," which became the company's motto.

By 1998 the Washington-based Solar Electric Light Company was capitalized (see chapter 5), and we could begin a multiyear process of pumping $1.2 million into SELCO-India to finance its growth. Harish and his team opened solar service centers (SSCs), which were scattered far and wide to serve the unserved and were usually located in small towns, preferably close to a rural bank branch, and which were all connected by computers. The headquarters in Bangalore knew exactly what was going on in any of the rural SSCs and could track inventory, accounts, customers, personnel, and sales.

Hundreds of SELCO billboards, called "hoardings" in India, and painted wall signs went up: "Make Your Home Bright; Use SELCO Solar Light." USAID funds paid for the production of SELCO television ads for the myriad cable operators to show. Business took off.

Our customers were mostly cash-crop farmers: organic spice growers who shipped their produce to Holland by air; workers on arica (betel) nut, sugar, and rubber plantations whose owners

SELCO-India "hoarding" (billboard) in Mangalore, India, advertising both solarelectric and solar hot water systems. *(Author)*

sponsored their employees' lighting systems; and growers of cashew nuts, coffee, coconuts, peanuts, vanilla, cardamom, black pepper, cloves, nutmeg, turmeric, ginger, and tamarind. Many of these farmers refused to use pesticides or chemical fertilizers, and their inherent sense of the rhythms of sustainability and the natural order of things rendered them keen to harness the sun for electricity. They understood.

Nothing was more rewarding for me during the twelve years I spent traveling to "the field" in all these countries than visiting Indian farmers who had recently purchased an SHS from SELCO. Surprisingly, many of these "poor" farm families lived in substantial houses, or bungalows, with tile roofs, three to six rooms, clean-swept porches, flower gardens, shade trees, and sun-drenched courtyards, where they could dry their crops.

The women benefited most, for their kitchens were dark, blackened with soot from cooking fires. We couldn't replace the cooking fuels (wood, coal, briquettes, kerosene), but at least the women could now see what they were cooking! (SELCO would later get into the efficient-cookstove business.) I would be offered flowered leis, glasses of coconut water, sometimes a Coca-Cola or strong coffee. It was rude to refuse, and going from house to house became an exercise in diplomacy.

The husband would take me from room to room, proudly turning on the lights and, if they had one, the black-and-white TV. Often he brought out their "retired" lighting systems, which consisted of small tin "bottle lamps" with unprotected wicks that created fire and health hazards beyond imagining. It was always rewarding when the family noted that their monthly loan payment to the bank was not much higher than their prior monthly expenditure on kerosene and candles—and their SHS could also run a TV, an added benefit for little additional cost.

Sometimes I'd be taken to visit a rich farmer, for whom SELCO had installed a government-subsidized solar water pump powered by a 900-kilowatt solar array, twenty times bigger than the standard SELCO household installation. This was a large contraption that was mounted in the courtyard on metal poles and could lift water one hundred feet. Farmers often found their solar-powered submersible pump delivered more water than they needed for their fields, so they had excess power that SELCO obligingly—for an additional price—channeled to the house for lighting. It was not uncommon for these farmers to have satellite dishes, color TVs, and lights in as many as a dozen rooms. So much for the "rural poor."

In the same category was a young "gentleman farmer" who cultivated his inherited land for his ailing father but who made his living trading stocks on the Bombay stock exchange using a solar-powered computer. This was in 1998! We drove along narrow back roads to his rambling farmhouse, far from anywhere in central Karnataka, following the telephone poles that brought a single line to the remote farm. He was ten kilometers from the nearest power line. (Telephones operate on a trickle current in the lines themselves; thus, Indians commonly have telephones but no electric power.) Closing the so-called digital divide would become a new business opportunity for SELCO as it later powered Internet kiosks and home PCs with appropriately sized PV systems.

Other customers included urban households and small businesses already connected to the power grid. Some people got fed up with the unremitting power outages and opted for a SELCO lighting system as a "backup"—and soon found themselves using the SELCO lights even when the utility power was working. One well-off customer in a small town in north Karnataka (he owned a car and a big two-story house) eagerly showed me his sheaf of electric bills, noting how they had decreased as he relied on the solar lights in lieu of grid power. What he saved on his utility bills he used to pay the bank loan. The utility connection was used mainly to run their refrigerators, which needed only half a day's worth of power to keep food cold. Rural households had no refrigerators, so it wasn't an issue.

There is in Dakshina Kannada (south Karnataka) a remarkable, self-sustaining community that few foreigners ever see. This is Dharmasthala, a mysterious and mystical place run by a powerful religious leader, one Dr. Veerendra Heggade. He represented the twenty-first generation of "temple managers" at a highly revered sacred Hindu shrine. Mr. Udupa and Harish had been telling me about this legendary figure for years because they wanted his support for a huge solar-electrification project in the communities surrounding "his" temple. I was told he was a "living god."

Besides his ancient Hindu temple, Dr. Heggade also managed an enormous rural-development organization called Shri Kshetra Dharmasthala Rural Development Project, or SKDRDP. Mr. Udupa was on the board of directors of SKDRDP. Dr. Heggade has a Ph.D. in agricultural and community development, and he had highly experienced rural development experts working for him.

In 1999 Dr. Heggade wrote in his annual report, "The biggest challenge of the new millennium is achieving development without harming the delicate balance of the planet earth. Struggle for food, shelter and clothing still remain our primary concerns. . . . Conservation of flora and fauna amidst rapid industrialization is forever a hard balancing act. This necessitates the adoption of a holistic approach to development. Sustainable Rural Development is the only answer to all these questions." He didn't want donor money because it mocked self-sufficiency.

He did, however, cajole the state government into providing him with a subsidy for solar electrification so that fifteen hundred families could afford to purchase SHSs from SELCO. The subsidy was paid directly to SELCO, allowing the company to discount its systems by that much. People got electricity, SELCO made a profit, and the government spent less than it would have subsidizing conventional electrification.

"Why couldn't this happen everywhere?" I later asked Mr. Udupa.

"Because there are many vested interests in India who wish to keep people poor," he explained. "So they can be controlled."

Well into SELCO-India's fifth year, I was taken to meet the illustrious Dr. Heggade at his compound deep in the forests of the Western Ghats, about thirty kilometers off the road that winds down to coastal Mangalore from the Deccan Plateau. Harish had explained how they had staged the fifteen-hundred-house project during the past year, employing additional technicians and hiring two cooks to feed them. They had a specially rented facility, where they stored the stacks of solar modules and hundreds of batteries that needed to be precharged by a large solar array before delivery to the householders. The region was so vast, with no maps or roads to speak of, that each family who had paid SKDRDP for the system had to come in person to pick it up and carry it home. Using some strange homing instinct (now they have GPS on their iPhones), our technicians would then follow up with a scheduled visit to install the system in these remote houses scattered through this dense jungle, where the hindrance to solar power was too much shade—so some trees had to go.

I went to Dharmasthala with Harish, and we arrived in a SELCO jeep, driving under the tall stone gate of Dharmasthala and entering one of the world's most unusual village centers. Here was Dr. Heggade's thousand-year-old temple to Shiva, with its complement of scantily clad priests. I paid it a visit to honor the small Hindu gods (Shiva, Vishnu, and Ganesh) deep within its hot, dark interior. Outside, I was blessed by one of the temple elephants, which lifted its trunk to gently touch my head.

Then, we visited the great marbled and columned dining hall, where up to thirty thousand pilgrims a day were fed by SKDRDP for no charge. I ate there and saw the operation myself, noting copper pots the size of Jacuzzis, filled with lentils (dhal) and rice, in the great kitchen. The tasty repast was served on palm leaves, with everyone seated on the vast marble floor in neat rows. This was an India few ever see.

I was taken to our quarters in the hundred-room guesthouse, owned by Heggade. After a wash and a rest, we walked to Heggade's favorite facility . . . his antique car museum! Smack in the middle of these jungles was an amazing collection of some thirty cars from another era—Rolls, Mercedes, Buicks, and Chevrolets once owned by maharajahs. They were all kept in running order, although they never went anywhere. Outside the museum, an added attraction for the pilgrims and tourists was a World War II Dakota transport plane. Don't ask me why or how it got there. The car museum was only surpassed by the enormous art and historical museum down the road, Heggade's very own Smithsonian.

Where was Heggade? "We will meet him after his daily audience," Harish said. "He is the law of this land, as well as its spiritual leader, and every day thousands of people line up to bring him their problems and disputes. He settles them with a few words of advice, which is like the law."

Sure enough, outside his dwelling, across from the temple, was a long line of supplicants waiting for a few personal moments with their living god. I was taken into the back of the audience hall to watch the last of the line present themselves to the master. He listened to each one, talked plainly to them in whatever language was necessary, dispensed justice, and cordially sent them on their way. He reminded me immediately of Dr. Ariyaratne of Sri Lanka, whom he in fact knew and revered.

Afterward, he went into the attached house to wash up, and we were ushered into his private quarters. He came out to greet Harish and me. Dr. Heggade settled comfortably into a chair, seemingly unrattled by shaking hands with over a thousand people that afternoon . . . as he did every afternoon. He was large, powerfully built, and had a wide moustache, a wider smile, and big sporty glasses. We were brought tea; a copy of *Modern Photography* lay on his coffee table. "Photography is a hobby of mine," he said.

Dr. Heggade talked matter-of-factly about his organization and about his other endeavors—his dental college, medical college, and engineering college up on the plateau. I'd noticed these modern complexes while we were driving down. He told us how he distrusted government but got along with the authorities nonetheless, all the while believing only self-reliance would save India, as Gandhi had preached. We talked about the challenges of running a socially responsible business, and he gave me a pamphlet of a lecture he'd given at his business college, entitled "Ethics in Business Management," in which he'd commented that "multi-national and transnational companies have completely reversed the value systems, ethical standards and social mores" in today's world. This was in 1999, before the multinationals really took over the world and before they began to dominate the Indian economy.

He quoted the Rig Veda, one of the Hindu scriptures—"Attempt not to obtain wealth through unjust and condemned ways." He claimed his temple was run according to modern management principles of "authority, responsibility, accountability, and transparency." This fellow would *never* be invited to lecture at the Harvard Business School.

The next day, we followed a maze of rough jeep tracks and walked jungle paths to visit a half dozen houses whose owners had purchased SHSs. It was time for flower bouquets again, and coconut juice and strong coffee and smiles and photos, and my shirt sticking to my skin while my Irish complexion got redder and redder in the tropical heat until I looked like a wet tomato. I was humbled by the effusive "thank-yous" from dozens of Indian farmers and their families, who sang the praises of SELCO for all to hear. Coconut juice straight from the husk never tasted better.

People were now buying a "SELCO," as they called our SHSs, validating our brand in a way we'd never imagined. In India, "SELCO" was becoming the generic noun for "solar electricity," as "Xerox" has become the verb for "photocopying." In 2002 SELCO-India moved into attractive new offices across the street from the headquarters of Wipro, India's largest software firm. I spoke at the dedication ceremony, which was attended by top Karnataka industrialists and officials, retired functionaries, and notable women leaders. SELCO's female managers turned out in bright silk saris. The rooftop party, beneath a colorful embroidered tent, continued into the night. The three-story headquarters building was entirely outlined in lights, like an extravagant outdoor Christmas decoration in America. (It has since moved, again, to larger offices nearby.) I was very proud that this young team of doers and believers had succeeded at something nearly everyone had said was impossible.

The World Bank now wanted to play with us. Their ambitious solar program, announced initially at the Pocantico workshop in 1995, was called the Photovoltaic Market Transformation Initiative, or PVMTI. It became our nemesis. Managed by intransigently difficult officials, PVMTI nearly killed SELCO, the parent company, and SELCO-India.

The $30 million PVMTI project was largely based on what World Bank contractors and consultants had learned from SELF and SELCO and a few other solar entrepreneurs. We cooperated for years with bank staff while this "initiative" was in preparation, having been told that the program would transform the solar markets and make the dream of "selling solar" that we'd discussed at Pocantico a reality. We were proud to have been able to influence this effort.

Then the years went by. Highly paid MBAs from McKinsey management consultants and various investment banks had gotten into the game and rewritten the rules to suit their own bottom lines. SELCO expected to get a couple of million dollars of the low-cost loan slated for India. Despite our "priority fast track" status, we didn't. Instead, the newly formed Shell Solar India took most of it. "Shell is a blue-chip company," the slick young MBA on PVMTI's London-based management team told me, "and SELCO is not." So much for a fund formed to help small entrepreneurs bring solar electricity to rural people. We hadn't expected to be competing with the world's second-largest oil company for these World Bank investment dollars.

In the end, however, SELCO managed to secure $1 million of low-cost financing from PVMTI, plus a "training and marketing" grant of $100,000. And they had finally agreed to make the loan in rupees instead of dollars.

All our solar success in India was, meanwhile, attracting attention, including from the Clinton White House, which notified me one day that the president would be traveling to India soon and would like to know what SELCO was doing. I provided stacks of information to the White House. Then, an invitation arrived addressed to Dr. Harish Hande, requesting his presence at the signing in Agra of the India-U.S. Clean Energy Agreement, which was intended to put $500 million into renewable energy in India. Harish was the only representative from the Indian or U.S. solar industry invited to the formal signing ceremony in the shadow of the Taj Mahal, and we all worried about what he would wear.

No clotheshorse, Harish usually wore sandals or scuffed shoes, no belt, and an open shirt, whatever the occasion. Indians are casual, except for affairs of state, when formality rules. Harish told me afterward that he had taken a seat in the small reception area (only about one hundred people were invited) next to Priyanka Chopra, the current Miss India. He told her to stick by him and he'd make sure she got to shake hands with Clinton. But he had an ulterior motive—he opined, correctly, that Clinton would come over to meet him if she were there, since no one could take their eyes off the lustrous Ms. Chopra. Harish figured he got in an extra couple of minutes with the president as Clinton gazed at her beauty while listening to him talk about solar. Ms. Chopra said nothing. She became famous a week later when she won the Miss World contest.

"So what did you wear?" I teased Harish over the clear, fiber-optic telephone line to India.

"Same thing as the president. Open shirt, casual slacks. Everyone else was in suits and ties." On TV I'd seen Clinton, with daughter Chelsea, at the Taj Mahal, where he'd been before the clean energy reception, and, sure enough, he was wearing a red sport shirt and slacks. Afterward, instead of hanging around with Ms. Chopra, Harish noticed a "sort of wallflower, all alone" at the reception and went over and talked to U.S. secretary of state Madeleine Albright for the rest of the event. Very noble.

SELCO-India took no Indian government contracts because its management was too honest: no corruption is tolerated. Personal kickbacks and under-the-table commissions are part of most government transactions with the private sector, but SELCO would not and still does not play along and thus has lost potential business. The Indian Renewable Energy Development Agency (IREDA) was the company's other nemesis in the early days. IREDA had hundreds of millions of dollars of World Bank funds to develop renewable energy, including wind, small hydro, solar hot water, biogas, bagasse (rice husks and sugar stalks to fuel power generation), and, of course, PV.

I met IREDA officials many times in their modern New Delhi offices and was promised the world. Harish and I attended their conferences and seminars, and I quickly learned that the entire operation was a smoke screen to protect a bureaucracy that had only two interests: its salaries and earning large management fees from the World Bank. The best way to look after a World Bank loan was not to spend or invest it, and the best way to protect your job was not to take risks. Do-nothing bureaucrats are never fired. I was told by educated observers that SELCO was getting none of this renewable energy money, even though we were seen worldwide as a visible success story, because we weren't offering the usual kickbacks. I credit SELCO and its incorruptible employees with helping to redefine business ethics in India.

One way to end your career at the World Bank is to highlight inefficiencies and corruption in a member state. India, to its credit, spends a lot of time rooting out endemic corruption; top government leaders from all parties are routinely thrown in the can, and the legal system nails lots of CEOs. A giant nationwide anticorruption crusade in 2011 led by Anna Hazare expressed the strong will of the Indian people to institute government reforms.

IREDA finally made a modest $100,000 low-cost loan to SELCO so we could sell SHSs at subsidized interest rates to a spice growers' co-op. The paperwork required was onerous beyond belief: half a foot high, a year to prepare.

The horror stories of dealing with the Indian government never ended. SELCO limited its interaction by having less and less to do with the bureaucrats in New Delhi. No energy minister, or minister of "new and renewable energies," ever visited a solar electrified village (that we know of). After a decade of successful solar lending by rural banks, this program was nearly brought to a halt in 2011 when the government attempted to satisfy rural energy markets by

Roadside flower stand enjoys nighttime operation thanks to bright solar-powered fluorescent light. *(SELCO-India)*

instituting complex interest subsidies that backfired, resulting in Syndicate Bank's abandoning their solar-loan program altogether. Other subsidy programs included a "per watt" rebate for solar home systems that companies were forced to pass on to their customers but that nearly bankrupted many solar businesses while they waited for the government to reimburse them, which could take years.

But before government meddling caused more harm than good in later years, SELCO became the beneficiary of an *international* financing package that leveraged $7 million for subsidized solar loans through Karnataka's rural banks. This was provided by the United Nations Environment Programme (UNEP), and cofunded by Ted Turner's United Nations Foundation. A young, no-nonsense UNEP program officer based in Paris visited me in Washington and Harish in India and then pushed the funding through. A letter addressed to me from UNEP arrived in 2004 with a surprising compliment: "You have done more in 12 years than create a great company. You've created an industry."

Tata BP Solar (now Tata Power) continued to supply SELCO with solar modules manufactured in its ever-growing factory north of Bangalore. The company's chairman, who himself had identified twenty-five thousand Indian villages that would never get electricity, was a big fan of Harish and SELCO's mission. Eventually, SELCO would "source" its solar panels from many different suppliers, some locally made, some imported once the government ended import duties on solar components. Mrs. Rao's Anand Electronics continued to supply its exclusive electronics and DC lights, while private-label deep-cycle batteries were outsourced and manufactured to the company's specs. SELCO opened a rural training institute at Manipal to train technicians. In 2011 SELCO partnered with a techie outfit, Simpa Networks, to allow customers to purchase solar home systems through pay-as-you-go mobile banking on their cell phones. Even if they don't have electricity, *everyone* in India has a cell phone, which they have to get charged in town.

A young Indian couple I met in Boulder, Colorado, where they both earned their Ph.D.s at my alma mater, CU, knew about SELCO and admired Harish's work on behalf of the poor. I strongly urged the husband to go back to his homeland and work for SELCO—they needed people like him. Anand Narayan took a well-paid job in IT while his wife completed her doctorate, and both could have stayed and earned big salaries in America. I was thrilled to learn that they had, indeed, returned to India and had both gotten jobs with SELCO. Anand was soon running SELCO-India's Incubation Labs, inventing new products, overseeing systems engineering, and directing technological innovation and strategic planning, while training young entrepreneurs in solar technologies.

Dr. Hande, as everyone called him, strongly believed in employing local youth: "Most of our employees are from the hinterlands. Local talent is our partner," he would tell *The Times of India* in 2011.

"They don't have a great education, but they understand the need for energy in a household. The technical competence of rural youngsters beats the theoretical knowledge of any IIT [Indian Institute of Technology] graduate." Himself a graduate of the IIT, he never believed universities turned out practical technicians or determined entrepreneurs.

The SELCO brand had become well known, indicating quality and reliability and providing "sun at your service" as the roadside "hoardings" proclaimed. By 2013 Harish and his people had sold and installed over 150,000 solar home-lighting systems in India, illuminating the lives of close to a million people. These customers were financed by over 400 *grameena* banks, including 150 that had been electrified by SELCO. Not all installations were small forty- or fifty-Wp household systems; many were multikilowatt systems for banks, universities, schools, temples, mosques, churches, and Christian seminaries. (There are 25 million Christians in South India who trace their faith back to St. Thomas, said to be buried in Chennai.) Solar water heaters became a popular product, saving people from having to burn fossil fuel or wood to heat water for bathing and washing. Solar water pumping became a substantial part of the

Indian technician installing twin solar modules atop rural home. *(SELCO-India)*

SELCO-India technicians and Karnataka farmer with solar-powered irrigation pump. *(SELCO-India)*

business. SELCO's sister firm, Anand Electronics, manufactured solar streetlights for purchase by local *panchyats* (village governments) and community organizations.

Through the years Harish maintained his devotion to his market: poor people, an immense, every-growing sector that could never be sated. He told not only the numerous reporters who came to interview him but also the many graduate student interns who came from around the world to learn from SELCO that "referring to the poor as a market is really a vulgar term." He said, "We have realized over time that even the poor are willing to pay for a service if the value they see in it is directly connected to an improvement in their quality of life." His credo is, "We need to treat the poor as partners, not as victims and beneficiaries."

American professors had exhaustively studied this vast lower economic tier since the 1990s, calling it "the bottom of the pyramid"—offering the largest untapped opportunity to sell goods and services in the world. SELF and SELCO were cited in early studies as examples of how to reach this enormous socioeconomic strata, especially in India. But corporations from Lever Brothers to Phillips to Coca-Cola (and, yes, Shell initially) saw only the extraction of profits, not social purpose or benefit. Also, multinational corporations operate *only* from the top down, and our approach to free-market small enterprise was entirely a bottom-up business model. This successful model, and SELCO's unique vision and values, attracted the notice of global institutions that were dedicated to triple-bottom-line economics (people, planet, profits).

Harish and SELCO started winning awards in the new century. First, it was Britain's coveted Ashden Award for Sustainability, presented to him in London by Prince Philip in 2005 (with a $50,000 check). Then, he won it again in 2007, and I received an e-mail of a photo of him getting this award from former vice president Al Gore, who gave Harish his cell phone number so he could keep him updated on SELCO's efforts combating greenhouse-gas emissions. Next came multiple complimentary invitations to attend the Clinton Global Initiative in New York, all expenses paid, where we would

Dr. Harish Hande with Vice President Al Gore at 2nd Ashden Awards Ceremony, London. *(Ashden Awards)*

meet each year (see chapter 10). He was invited to "renewable energy" conferences around the world. National newspapers and magazines began running full-page articles about him. None of this ever went to his head, but he did change his sartorial habits, which never extended beyond the aforementioned plain shirt and beltless slacks: he began wearing the traditional Indian sleeveless tunic and white pantaloons. True to form, he has *never* put on a Western jacket or tie.

He was following the dress code of another well-known solar entrepreneur from North India, Bunker Roy, who always appeared at global events looking like an Indian politician. They only wear native dress, never Western suits, as the businessmen of India do. Gandhi always wore a native dhoti, and it worked for him. Interestingly enough, no urban Indians dress this way. Women may adhere to their saris, but men in India at all economic levels dress "Western-casual."

What was important was that Dr. Harish Hande and the company we so lovingly and adventurously cofounded in 1995 was finally getting recognized. When we closed down the parent company in Washington in 2003, I resigned as chairman, for the company was now in good local hands. SELCO was restructured as an entirely Indian company with direct foreign investment rather than a subsidiary of an American firm. New investors from the United States and Europe provided fresh capital. Harish no longer reported to me but to a new board that believed in SELCO's mission. SELCO had succeeded in its ten-year battle to bring affordable electricity to those who otherwise would never have had it. As Gandhi once said, "First they ignore you, then they ridicule you, then they fight you, and then you win."

In 2008, Kurt Schwab, the founder of the World Economic Forum, invited Harish, all expenses paid, to come to Davos, Switzerland, and speak about the SELCO model of sustainable enterprise. There he had his say before global capitalists, world leaders, CEOs, other social entrepreneurs, and the media. He would explain, "Our company has the difficult task of balancing social objectives with commercial viability; we never make spectacular profits, but just enough to keep us afloat."

He has been invited back to Davos every year. At Davos in 2012, where, according to *The New York Times,* they were "debating capitalism's future," he took on the Harvard economics professor and author Michael Porter, who had said the only measure of effectively doing social good was making money. Well, of course, a so-called social enterprise must make money, but Harish was sick of this capitalist mantra. He wrote from Davos in *The Huffington Post* in January:

> [Mr. Porter] said capitalism is the only way of measuring sustainability. I completely disagree.
>
> Capitalism in isolation is a perfect recipe for social disaster.... We need to look at social sustainability in a more holistic and long-term way.... History shows that stable societies were not built on capitalism. Successful civilizations were built on the basis of balancing social and financial sustainability. And today... we have to add another dimension: environmental sustainability.... This balancing act cannot be done by corporations alone—but only with the partnership of government and civil society with equal responsibilities.

He was echoing what I had earlier learned the hard way at our parent company, which historian Cyril Aydon (*The Story of Man*), while writing about the "huge potential harm that uncontrolled profit-seeking carries with it," described thus:

> A socially responsible business—like a flying pig—is a contraction in terms. Pigs are not designed to fly, and privately owned business corporations are not designed to serve the public good. This is not to characterize them as public enemies, merely to recognize them for what they are: devices to enable their shareholders to maximize their earnings, and their senior executives to maximize their power *and* their earnings.

This was the dilemma all social entrepreneurs faced. Fortunately, their prospects, and SELCO's, were not quite as bleak as Mr. Aydon expresses above, since they are all bottom-up private companies utilizing the corporate model—financial investment and stock ownership—but with patient capital for the most part, and directors who themselves are trying to rewrite the rules of the otherwise irreconcilable conflict between global capital and small-to-medium enterprises. And most "social enterprises" are not public companies so are not influenced by share prices.

The point was not to reform large corporations, which perhaps can't be reformed, but to replicate the SELCOs of the world and encourage entrepreneurs to tackle problems from food to energy to health care and clean water—and jobs, which many are doing around the world right now. Scaling up successful business models like SELCO can best be achieved by replicating that model: rather a thousand SELCOs than one General Electric to serve the have-nots of the world. Clearly, small companies are better at providing distributed and decentralized energy than big ones (see Amory Lovins's *Small Is Profitable*).

Let me put all this in my own less politic words: I *hate* the corrupt finance capitalism that brought down America's economy in 2008, but, as I said earlier, I *love* business, commerce, free enterprise, risk taking, fair trade, and truly free markets . . . and *profits*. This is a clear distinction that few in Davos, no business schools, and not many corporate and financial leaders are willing or able to grasp. They mostly believe that if you criticize corporations and unsustainable and corrupt business practices, you are against capitalism, and this must make you a socialist or Communist—such is the juvenile level of debate about economic justice in America. Boston hedge fund manager Jeremy Grantham, who manages a $97 billion investment portfolio, recently wrote that "capitalism threatens our existence" by depleting resources and that "growth at any cost" is a recipe for planetary suicide. He added, "Washington is becoming a corporate subsidiary." The Tea Party in America would, no doubt, call this multibillionaire a Communist.

Known as one of India's leading social entrepreneurs, Harish was named by *India Today* magazine in 2008 as one of the fifty pioneers of change in India. In November 2011 he was invited to participate in President Obama's Entrepreneurs Roundtable in Mumbai, where he was the only Indian-solar-company representative. "Entrepreneurship is the only remedy for job creation," these young business leaders told the president. Earlier in the year, Harish had flown to Manila to accept the lofty Magsaysay Award, known as the "Asian Nobel Prize." Despite all this notoriety, he remains humble. No awards or plaques or photos of him grace his simple office in Bangalore; there is no vanity wall. "He never talks of what he has achieved over the past eighteen years," a member of the SELCO team told reporters.

India may indeed be the crucible where new forms of business governance and social enterprise emerge. More importantly, India is a world beacon of religious tolerance—for the most part. OK, Muslims slaughter Hindus on the slightest provocation, and Hindus respond with massacres of their own, but the majority of India's 150 million Muslims and one-billion-plus Hindus get along, comingling in every town.

The country welcomed the Tibetans when they came over the Himalayas in the 1950s, when the Dalai Lama and eighty-five thousand refugees were fleeing Chinese aggression. As an aside, it is interesting to compare, as I have been able to do, two next-door nations with populations of well over 1.3 billion people each: one deeply spiritual, where religion of one kind or another holds sway, the other nearly pagan and almost entirely without religious values or a spiritual base of any kind. I'll take India any day, for China is heartless and mean by comparison. But China is today the far richer and more dynamic country.

The Tibetan refugees in India live an isolated existence. The government settled them in ten states, including five settlements in Karnataka, which now hosts the country's largest Tibetan communities. Hindus seem to revere these newcomers who are spiritual seekers like themselves. There is little possibility they will ever go back to China, and it is equally unlikely they will ever integrate into India. What will become of them is anyone's guess.

The Dalai Lama with SELCO-India's Suresh Salvagi (left) and Nityand Mukherjee celebrating Tibetan Buddhist temple's new solar lighting system installed by SELCO in Mysore State. *(SELCO)*

SELF's Bob Freling had met the Dalai Lama in Washington, D.C., and Bob garnered financial support in the United States from people, especially his board members Steven and Mary Swig, interested in helping the refugees in India. Bob contacted Harish, who visited the Tibetan settlements in Karnataka to assess their power needs; it turns out they had no power. SELF, together with the Tibetan monks themselves, funded the Solar Lights in Tibetan Settlements program. Harish sent his best technicians to live in the community for nearly a year and oversee the project that brought street lighting, illumination for monks' cells, solar water heating, lighting for hundreds of households, and solar power for the SOS Tibetan Children's Village. Solar lighting was installed throughout a new multiacre temple complex and monastery, which included the Dalai Lama's private quarters and bedroom.

When the Dalai Lama visited South India for the dedication, he met Bob and his wife, Chaya, thanked him for SELF's support, and then sought out SELCO's project managers and thanked them enthusiastically for harnessing the sun's energy and bringing electricity to the Tibetan temple and community. He drew them close, then took their hands in a firm grip and smiled for a photo that now graces every SELCO office in India, often mounted right next to the ubiquitous garish calendars featuring all the Hindu gods, who don't seem to mind.

By mid-decade in the new millennium, the Congress Party returned to power with a strong rural vote, indicating India's 8 percent growth rate—as India opened up and foreign capital and goods poured in—was having little economic impact on 700 million peasants. In some ways, globalization was hurting them as factories were closing, unable to compete with cheaper Chinese goods, including imported silk. Farmers were suffering from the influx of imported foodstuffs, and many committed suicide. Despite all the talk of high-tech and the growing middle class, only a million Indians out of a billion (one tenth of one percent) worked in IT. And even today, only a tiny percentage of the population own computers. One of the Congress Party's election promises had been "free electricity for rural areas." Electric power was, as always, the paramount issue, but it can never be free. Even "free power from the sun" costs something.

Over 400 million people are still without electric lights in India, let alone power to run a TV or a computer, and the energy crisis remains. These people enjoy a lifetime electricity blackout, unphased by the "world's worst blackout" in July 2012, which cut power to 620 million Indians normally supplied by the central grid—nearly one tenth the planet's population! Now, India is beginning to take solar electricity seriously. Whether the top-down, large-scale government-corporate approach will bring power to the poor remains unclear. Harish doesn't think so.

In 2010 the Indian government under the progressive leadership of Prime Minister Manmohan Singh embarked on an aggressive plan—the National Solar Mission—to increase the use of solar power to 20,000 megawatts by 2022. The goal was based on large subsidies and the 70 percent drop in global PV panel prices. This top-down versus bottom-up, centralized versus decentralized solar policy went counter to the distributed-energy model pioneered by SELCO-India. Harish e-mailed me:

> Centralized solar is socially very expensive and unsustainable. The incentives have proven that subsidies are for the rich—investors in large solar plants—and the poor are left with scraping the bottom of the pan.

Nevertheless, all power from the sun is good, and the good news here was that centrally generated solar electricity was now competitive with diesel power generation in daytime hours, even if it didn't reach the poor. India has no oil and imports what it uses in its thousands of diesel power plants and small industrial-generator sets at a cost of 65 cents a kilowatt hour! India also burns dirty coal, which is expensive to transport, compounding greenhouse-gas emissions. So even central solar is good, as long as the economics work and the private sector is willing to take the risk. Let all solar, big and small, play by the same rules: if it pays, it plays.

One player is Azure Power in Khadoda, Gujarat, where thirty-six thousand solar panels on sixty-three acres feed power into the national grid. With India opening its state-owned power-generating industry to private operators, the solar boom was on: Chinese manufacturers (Suntech, Yingli) and American manufacturers like First Solar rushed in to build central PV plants; CSP (concentrated solar power, or "solar thermal") developers started planning fifty-plus-megawatt generating plants. Indian PV manufacturers geared up for this huge domestic market. In early 2012, Rajasthan was accepting bids for two hundred megawatts of solar power.

A leading Mumbai bank, whose chairman we had educated about solar a decade before, financed a 15-megawatt solar farm in Gujarat. This was a small part of a 600-megawatt "solar park" in that state, dedicated in 2012, which will save 900,000 tons of coal and reduce CO_2 by eight million tons during its working life. This is the largest collection of giant solar arrays in the world, in a single "solar park" in crowded India. New financing mechanisms such as solar-backed securities were being looked at. India is expecting to install 3,000 megawatts of solar annually from here on. The government buys all this independently produced solar power and, by bidding for the lowest cost of power generation, creates competition among solar companies, domestic and foreign. Everybody wins, except the 400 million people with no power, since they will see none of it as they are beyond the reach of the notoriously inefficient national grid. But at least they are getting solar power, thanks to companies like SELCO.

Quoted in *The New York Times* the day after the catastrophic 2012 North India blackout, Harish said wryly, "Hopefully, the 400 million Indians who do not have electricity today, *because they never had any,* will show the rest of India how to behave in the future."

The best idea I've heard for distributed solar in India on a massive scale is from an American of Indian descent, Jigar Shah, one of America's biggest PV evangelists and founder, with Virgin Airways CEO Richard Branson, of the Carbon War Room. Shah plans to solar-electrify tens of thousands of India's 300,000 diesel-powered cell phone towers. One cell phone tower operator with 32,000 towers has already replaced many diesel generators with solar, saving tens of thousands of liters of diesel fuel a year. Indian billionaire Sunil Mittal, owner of India's largest cell phone company, Bharti Airtel, has started swapping backup diesel for photovoltaics. Ironically, it was this global telecom market that kept many of the world's solar manufacturers afloat in the 1980s and 1990s. Cell phones have created a whole new and unanticipated power demand, just as cable TV did, and India's notoriously unreliable power grid can't cope; hence India's reliance on distributed diesel power. But now distributed solar is stepping in. By the time you read this, India will have truly entered the solar age.

Meanwhile, SELCO Solar Light Pvt. Ltd. (its official name) pays little attention to all the Big Solar Plans coming out of New Delhi and state governments because Harish knows that very little if any of this new grid-fed solar power will serve the poor and unelectrified. So SELCO continues to grow, with 320 employees and 36 regional energy centers in five states. The company plans to expand into more states and continues to train entrepreneurs nationwide. Since 1995, SELCO estimates it has reduced carbon emissions by 35,000 tons. That's a lot of kerosene!

If Indians who can't connect to the national grid want to buy electricity, SELCO and dozens of other "solarpreneurs" will be there, providing "sun at your service."

I didn't know it at the time, but what I'd learned in the land of the sacred sun would prepare me for a whole new adventure in the solar business as I was inspired to replicate the SELCO-India model back home in America (see chapter 11). This would come later. In the meantime, the learning curve would continue, and my solar odyssey would include Sri Lanka, Vietnam, Africa, and China.

7

SUNSHINE AND SERENDIPITY: 1991–2000

Serendipity, as you'll recall from chapter 2, brought me back to Sri Lanka, the teardrop island nation at the tip of India, which I had first visited in 1971. In our late twenties, my first wife and I set off round the world and spent a wonderful month together on the fantasy island of Ceylon, visiting all the tourist sites with a hired driver: the ancient cities of Polonnaruwa, Anuradhapura, Sigiriya; the beaches; the temples; the tea estates. We lived like sahibs at the Queen's Hotel in Kandy, at the government guesthouses set in lush gardens with monkeys in the shade trees, and at the historic Galle Face in Colombo, where we were the only guests. The Galle Face was a place to which I was serendipitously to return exactly twenty years later.

In March 1991, Lalith Gunaratne pulled up in front of the now-refurbished Galle Face Hotel on Galle Road (the coast route to Galle, the Portuguese and Dutch fort city in the south). Built in 1862 in the grand British colonial style, it is one of the oldest hotels in the world in continuous operation. It anchors the Galle Face itself, a huge expanse of open grass south of the old parliament building between the Indian Ocean and Galle Road. Its rooms are large enough to fly small aircraft in, and ocean breezes make it tolerable when the feeble air-conditioning doesn't. Sri Lanka is seldom as hot as Washington, D.C., but its extreme humidity can drive you mad. You can come to terms with it on the long, open veranda beneath a dozen ceiling fans, where white-robed waiters with brass number tags solicitously bring you fresh lime sodas. Several of the waiters I remembered from 1971 were still there in 1991, and one even remembered me. All our solar adventures in Sri Lanka were launched from the

Galle Face veranda, including SELCO-Sri Lanka, which I'll get to momentarily.

Lalith jumped out of his red Toyota sedan and came around to greet me on the steps. Built like a wrestler, dark-complected, with a full beard, he looked like someone out of a Sinbad movie. He was stylishly dressed and spoke his Sri Lankan English with a lilting accent tempered by his upbringing in Toronto.

"Welcome to paradise!" he exclaimed. "Do you like old car rallies? Samantha and I want to take you with us to the Mercedes-Benz rally in Victoria Park."

"Sure," I said, surprised. So here I was on my first afternoon back in Ceylon, now renamed Sri Lanka, and I was going to an upscale Sunday-afternoon function with the cream of Colombo society to look at classic Mercedes-Benzes. It turns out there are a lot of them in Sri Lanka! I once had a used Mercedes, so I could appreciate what I was seeing.

Lalith owned a small, pioneering, solar PV company, which he and two of his fellow Canadian Sri Lankans had started five years before, deciding after college to return to the land their parents had left. Lalith had originally studied "nuclear plant management" at a Canadian university. I asked him later why he switched to solar. He replied, "Because I woke up one night in a sweat with the realization that splitting the atom is an unnatural act." He believed atoms, as the building blocks of the universe, should remain whole. He had formed Suntec with family investors and a big loan from the Development Finance Corporation of Ceylon. Lalith wanted to improve the lives of rural people with solar electricity.

Thanks to the miracle of fax machines, I had been in touch with Lalith regarding Suntec's intention to "sell electricity at the customer's doorstep." I wrote to him that SELF was promoting such efforts wherever we could, and that we could raise money for such projects; we'd already been promised a large sum from the W. Alton Jones Foundation to launch a solar project with Sarvodaya, as recounted in chapter 2. We would need solar home-lighting systems, and I decided we might as well buy them locally, if they worked.

Lalith and his friends had imported equipment from Canada to manufacture their own photovoltaic solar modules. They imported the cells from Japan, which then had to be cut, "tabbed," and laminated into 36-Wp modules. Theirs was, indeed, a daring enterprise, since they did not have the volume of sales to warrant local manufacture. It would have been cheaper to import finished modules. They also had problems with unreliable DC light fixtures and their inverter ballasts, which they also made, often with what they didn't know at the time were inferior parts from the local market. Nonetheless, Suntec supplied SELF's two solar projects in Sri Lanka over the years, working jointly with Sarvodaya's technicians—after helping to train them. Through these programs we electrified some thousand households, plus eighty Buddhist temples. Everyone, even the temples, had to sign up to pay on installment credit through one of several credit mechanisms SELF launched. Lalith and his two partners, Viren Perera and Pradip Jayewardene (grandson of the former prime minister, Junius Jayewardene), were three of the most extraordinary people I had ever been privileged to meet, all immensely dedicated to this wondrous technology that could bring happiness to the rural people of this magical isle.

Suntec had a good run (like SELF in Sri Lanka), and by 2000 the partners had sold it to Shell Renewables, which renamed it Shell Solar Lanka Ltd. Lalith became an international consultant on rural energy, and we continued to cross paths everywhere—at solar energy conferences in Washington, Morocco, Switzerland, China, Vietnam, Amsterdam, and Germany. Pradip became Shell Solar Lanka's CEO. Today Lalith lives in Ottawa with his family.

Sri Lanka, with its small but dense population of 21 million, was where I came to understand and to know the rural people of the two-thirds world—before India, before China, before Vietnam. Here, I was able to enter this world, their world, the world in which half of humanity lives. It is a world of small farms; of simple, small houses with thatch or tile roofs; of families usually too large to fit in the usual four rooms. Often, these rooms have no furniture and people sleep on floor mats. But it is not necessarily a world of "rural poverty," and the words "rural" and "poverty," I learned, don't necessarily go together. In fact, according to the United Nations, there are more urban poor than rural poor in the two-thirds world.

The two-thirds world consists mostly of proud, self-sustaining peasant farmers. There are 700 million of them in India, 800 million in China, maybe 400 million in Africa, at least 200 million in Latin America, and another 300 million in the Middle East, Southeast Asia, western Asia, eastern Europe, and Russia. Their lives have not changed much in hundreds if not thousands of years. They mostly own or lease their farms now, since both India and China instituted vast land reforms, so at least most are not serfs or sharecroppers. They are the majority population of the two-thirds world, constituting at least half the four billion people in the developing world who make up two-thirds of the planet's total population of seven billion. And half the four billion have no electricity; half of all peasant farmers have no electricity. This was the world I came to know intimately, or as intimately as a foreigner can without actually living there.

Until Lalith Gunaratne took me inside a peasant farmer's house, I had never been in one (except in the Dominican Republic). Now, I spent my days in a dozen countries paying endless visits to rural farmers, entering their houses; meeting their families; photographing their smiling, guileless children; patting their wary dogs; and taking tea in what were sometimes no more than wattle-and-daub one-room shacks. (I could have titled my first book *Many Cups of Tea* and maybe had a bestseller!) Many homes were more than shacks, and even Lalith never ceased to be amazed at how relatively well-off so many rural people actually were, living on single cash crops, a small garden, and maybe some livestock. They often had solidly built houses with hardwood trim, proper windows and doors, red tile roofs, ceramic floors, handcrafted furniture, running water (hand pumped and gravity fed), and numerous well-furnished bedrooms. But they had no electricity and no hope of getting any until solar power arrived.

"People in Colombo have no idea people out here live like this. All they think of is the 'rural poor.' Most of our bankers and industrialists have never visited rural Sri Lanka," Lalith lamented during one of our many rural excursions. "They send their kids to school at Berkeley and go on holidays to Europe, but they've never driven across the island to Trincomalee."

This is a world of unspoiled beauty, of peace and quiet, of country rhythms and timeless cycles. There are the quick twilights, the early dawns and clear mornings, and the hazy, lazy afternoons when you only want to rest under a tamarind tree after another noontime platter of red-hot "rice and curry" served on an aluminum plate. The farmers are invariably bare-chested, dressed only in a long, colorful, plaid or batik sarong. The women have simple workday saris, while the children are in school uniforms or casual but neat outfits: girls in frilly bright satin dresses, the boys in creased and pleated shorts with white short-sleeve shirts. No one wears shoes, but they have them for going to town.

People in this world are uncomplicated, often uneducated but not illiterate, and neither naive nor without their passionate loves and hates as they struggle against fate and nature. This was richly portrayed in *The Village in the Jungle*, a classic 1913 novel by Leonard Woolf (Virginia Woolf's husband), which he wrote after seven years in the Ceylon Civil Service. "The villagers belonged to the goiya caste, which is the caste of cultivators. If you had asked them what their occupation was, they would have replied 'the cultivation of rice,'" Woolf wrote. "The spirit of the jungle is in the village, and in the people who live in it. They are simple, sullen, silent men. In their faces you can see plainly the fear and hardship of their lives."

The hardship for so many of the rural majority of the two-thirds world is caused by climate and uncertain weather cycles. No rain means no rice, and they revert to slash-and-burn agriculture, cultivating *chenas* (clearings in the jungle); as Woolf noted, "So sterile is the earth, that a chena, burnt and sown for one year, will yield no crop again for ten years." In Sri Lanka, no rain also means no electricity, for every bit generated on the island is from "renewable" hydropower, which doesn't always renew, especially in these days of severe climate change. But in 1991 that didn't concern the 40 percent of the population, all of it rural, that had no power anyway. No, they were concerned with eating and with feeding a growing population on limited lands in an equatorial climate where soils are notoriously poor, where jungles grow and crops do not.

But despite the fragile existence of this ancient rural life, what I saw was "sustainability" in action, a work ethic almost unknown in the urban world, and a self-sufficiency that is unimaginable to most of us. (Lalith and I believed we could help make them SELF-sufficient in electricity, too.) How did the best-off among them pay for their nice houses, furnishings, motorbikes? Usually someone worked in town or in Colombo or abroad. Many young mothers worked as maids in the Gulf States, many a husband in Saudi Arabia or Oman. Rural life wasn't as bucolic and content as it sometimes seemed, because the people had very little cash; many used a third of their income to buy kerosene for lighting and cooking.

Rural life nearly came undone in Sri Lanka in the 1980s, when a Singhalese Marxist peasant movement, the JVP, emerged to challenge the state. Not unlike the Khmer Rouge, or the Shining Path in Peru, the JVP was led by an ardent "rural revolutionary" who thought the peasants in Sri Lanka were getting short shrift. This is the other and not so pretty side of the rural two-thirds world. Because people have little access to information, with few newspapers—and no TV because of no electricity—they begin to feel isolated and left behind by a world they can't be part of and don't understand. Young people in these rural areas are easily mobilized by a charismatic leader—an old story, perfected by Mao Tse-tung, whose uprising succeeded.

The JVP's uprising didn't succeed in Sri Lanka. By 1989 its revolutionary leader had been captured and killed by security forces. Global terrorism expert Rohan Gunaratna told me over lunch on the Galle Face veranda how this organization nearly shut down the country; worse was the government's response. The story remains nearly unknown in the West, but to root out the JVP, government forces are said to have arrested and killed thousands of teenage boys—no one knows exactly how many—from villages throughout Sri Lanka. Whole classes of boys were found beheaded on their school grounds. The government felt it had no choice. Rohan set me straight about those dark years, a period of Sri Lankan history no one wanted to talk about. (*Anil's Ghost*, a novel by Sri Lankan–born Michael Ondaatje, author of *The English Patient*, deals with the aftermath of those years.)

This shows why rural development is so important. It is critical to bring some basic services to the people living on the land. They will be better served by staying on the land, as opposed to joining a peasant revolution or moving to a city, where they will end up in squatter shacks and in slums. SELF's funders shared this view, and later we found investors who saw delivering solar power services to rural areas a signal opportunity for "responsible capitalism," and not just a social mission.

Others saw it as sound international policy: electricity as an entitlement. One member of the European Commission told me in Munich, in terse terms, "If we don't bring electricity to them there, they'll all come here." And coming they are. Sri Lankans, Bangladeshis, Afghans, Pakistanis, Indonesians, and North Africans are flooding into Italy, Spain, and Greece. They regularly wash up on the beaches of the Mediterranean, dead or dying.

The rural people I came to know as I was invited into their humble homes were decent, hopeful, generally happy, friendly, outgoing, and extremely capable at cultivating whatever crop they had chosen to grow. Development experts claimed they needed clean water, but they all had wells or access to good village cisterns. The donors' consultants said they needed medical care, not electric lights or TVs, but they all had access to numerous rural clinics. The international agency officials said they needed housing first, but nearly everyone I saw already had a nice house, with a neatly swept courtyard, flower gardens blooming profusely, shading palms, and food often dropping from the trees.

What they *didn't* have was electricity. They couldn't afford batteries to run radios; they tried to watch small 12-volt black-and-white TVs by charging car batteries, and recharging them was a common local enterprise—the batteries were carried on bicycles to the nearest town, a big pain. And their children, whom they cared about

Author helps Sri Lanka child turn on an electric light switch for the first time. *(Author)*

most of all, could not study well by kerosene lamplight. Their kids did not wish to emulate Abe Lincoln, not when they knew something better existed, which Abe didn't have: electric light.

For the next ten years I commuted to Sri Lanka from Washington, D.C., as we sought to bring solar rural electrification to this ill-fated if beautiful island. Unserendipitously, the year after the government put down the JVP insurgency, the Tamil revolt erupted. The Liberation Tigers of Tamil Eelam (LTTE) sought self-rule for the north end of the island and the city of Jaffna. LTTE's leader was the self-declared revolutionary Prabhakaran, a monster who had ordered 220 suicide attacks, outdoing the Palestinians, and predating the Iraqis and Pakistanis. I scribbled in my diary in 2000, "Prabhakaran is the most ruthless and violent terrorist running loose on the planet. He makes Bin Laden and his terrorists look like rank amateurs." Bin Laden would outperform Prabhakaran a year later, and Americans would know what it is like to live in a state of terror after 9/11/2001.

Over the decade I worked in Sri Lanka, the violence got worse and worse, but it seemed to have no effect on the good life in Colombo, on our rural projects or our operations, or on our ability to travel around most of the country. The war went on like an argument in the next room to which no one paid attention, except the rural families of the 64,000 government soldiers, Tamil Tigers, and civilians who had died. The news media barely paid attention to a war that consumed thousands of young soldiers every year; reporters were not allowed near the battles. Governments came and went; one prime minister, Premadasa, along with twenty-four of his aides, was blown up by female "suiciders" at an election rally (that democracy works at all in Sri Lanka is a miracle). Tiger terrorists blew up the State Bank in 1999, killing hundreds, then set off a huge truck bomb in the Fort, the financial center of sprawling Colombo, and blew out all the windows of the Intercontinental Hotel, the Hilton, Le Meridien, and the two thirty-nine-story towers of the brand-new International Trade Center.

Prabhakaran's Tigers had nothing to do with redressing rural poverty; it was a vicious thug's personal fight, nothing else, hiding under the guise of demanding ethnic rights. Tamils had been in Sri Lanka for a thousand years and were brought in from South India by the Singhala kings, whose people didn't like to work. They comprised half the population of Colombo, owned most of the large businesses and industries, and had a determination that sometimes seemed lacking in the more laid-back Singhalese. A famous Tamil family owned the Galle Face Hotel (maybe that's why I always felt safe there). But Tamils in Colombo had little or nothing to do with the Tigers. Neither did the Tamil tea pickers in the hills, brought in by the British; they didn't support what foreign supporters claimed was an ethnic uprising, and they weren't violent. Most good citizens of Sri Lanka recognized the LTTE for what it was: a demented, twisted, and deadly terrorist organization built on the backs, and lives, of brainwashed youth.

The Norwegians were everywhere, trying to negotiate a peace accord as they had done with the PLO and Israel, but they didn't have much luck. I sat next to the Norwegian ambassador on one flight out to Europe, and he told me he'd just come from a meeting with Prabhakaran (I knew, and asked him about it, for it was in the papers). He said they were ready for peace. I thought him naive. Two weeks later, in 2001, Tigers blew up nine jet aircraft on the runway at Bandaranaike International Airport, destroying military and civilian planes, including two of Air Lanka's brand-new Airbus 340s. Twenty Sri Lankans also died in the blasts, but the Tiger sapper squads took care not to kill a single tourist (unlike the mad Islamists in Egypt, who the same year had gunned down fifty-nine tourists at Queen Hatshepsut's Temple at Luxor). Tourists weren't really at risk in Sri Lanka, despite State Department warnings that kept virtually all Americans away (in ten years I never met a single American tourist in one of the most glorious tourist destinations on the planet). The European package tours continued to arrive from Leeds, Sheffield, Düsseldorf, and Amsterdam, the tourists ready for sun and beach and clueless that there was even a war going on. For them, Sri Lanka was nothing more than a cheap beach. I always felt safer in Sri Lanka than in Washington, D.C.

According to an article by Bill Miller and Thomas Lippman in the March 14, 1999, *Washington Post*—and corroborated by the records of the U.S. Court of Appeals, D.C. Circuit (97-1948)—the Tamil Tigers, the group Prabhakaran founded and led, was represented by former Carter administration attorney general Ramsey Clark in their effort to overturn the U.S. State Department's designation as a terrorist organization! This is an example of how little Americans understand the realities of the two-thirds world.

Priyantha Wijesooriya, whom I'll call PW hereafter, was a graduate student from the University of Massachusetts, Lowell, who had introduced me to Harish Hande. He also led me into business in Sri Lanka. He came to the veranda of the Galle Face one day in 1996 with his wife and daughter and brought me a business plan for a new venture he called RESCO, for Renewable Energy Service Company.

PW had just finished his Ph.D. in solar engineering (like Harish), with a masters from the London School of Economics, and so he was certainly qualified to undertake this enterprise. I had worked previously with him on pilot projects, such as the solar electrification of Morapatawa village, where he had raised sustainability to a high art. This was a very poor community of cashew growers north of Puttalam on the northwest coast, truly a village in the jungle, accessible only by five kilometers of dirt track. Here we had installed 150 Suntec SHSs and set up a local revolving fund with village management, and PW had even opened a solar-powered "jungle workshop," where technicians he trained began assembling and soldering together their own charge controllers and luminaires (DC electronic light fixtures). They also learned to install and maintain the systems. This became one of our truest models of SELF-sufficiency, visited by endless delegations of consultants, rural aid officials, and graduate students. Year after year the local "solar committee" diligently collected the monthly installment payments from each family, oversaw the maintenance of the systems, and reported regularly to PW in Colombo. With the funds collected, they began to make more loans to more families to buy SHSs.

PW was a strong force of muscle, will, and visionary brilliance, like Lalith, and I introduced the two of them one day early on, for they would have to work together. They became friends but remained in their separate spheres: Lalith's Suntec manufactured and supplied SHSs, while PW sold and installed them through our SELF pilot projects. He realized that these philanthropic projects were not enough to change the rural landscape from darkness to light.

I wasn't quite ready to make the transition to a commercial business, even though we had already done so in India. I concentrated my efforts on securing some large infusions of foundation funds from the United States for our Sarvodaya project. Dr. Ariyaratne, the most famous person in Sri Lanka (he declined many offers to run for prime minister), had decided to put the full force of his huge NGO behind the SELF/Sarvodaya solar project.

Lal Fernando, the chief of Sarvodaya Rural Technical Services, was named the solar program manager. Lal was a dedicated and professional engineer with a decade's experience handling Sarvodaya's rural water projects. He decided to take on solar electrification as well. A devout Catholic, he nonetheless strongly believed in Dr. Ari's concept of "right livelihood for a full-engagement society" and Sarvodaya's Buddhist ideal of "universal awakening," and he worked tirelessly for this great "people's movement," crisscrossing the highways and byways of rural Sri Lanka in his Toyota double-cab pickup.

Lal introduced me to a former banker who now headed up the Sarvodaya Economic Enterprises Development Services (SEEDS), the huge island-wide credit facility that we would rely on for many years to provide credit to rural householders so they could purchase solar lighting systems. SEEDS' banking division was providing microcredit to the poor long before it became fashionable in donor circles.

By the late 1990s, SELF, Suntec, Sarvodaya, and SEEDS—and PW's independent RESCO—were cooperating to bring solar lighting to some two thousand families island wide. Sri Lanka was becoming the world's crucible for solar rural electrification.

After Anil Cabraal and I led a World Bank mission to Sri Lanka in 1993, Anil, who was the Sri Lankan–born solar energy expert at the bank, and I convinced the bank's Asia Alternative Energy Unit to propose a $30 million Sri Lankan Energy Services Delivery Project (ESD) to the government of Sri Lanka and the World Bank board of directors. It only took six years from our 1993 fact-finding trip to get the project approved!

PW, or Dr. Wijesooriya, as his employees called him, wasn't waiting. He launched RESCO–Sri Lanka as a semipublic company, approved by the country's board of investment, with funding from a Dutch development bank, the People's Bank of Ceylon (the largest government-owned bank in Sri Lanka), and my newly formed Solar Electric Light Company, which took, initially, a minority share. In the paneled penthouse offices of the People's Bank chairman, overlooking Colombo, we sat around a huge table signing endless documents, shareholder agreements, bylaws, certifications, and various government papers officially creating the company in October 1998. I was named chairman, and PW was chief executive.

As best I could, I was now overseeing two commercial companies in two countries, inventing this new reality, which consisted of cash flows, balance sheets, income statements, loans, debt, interest rates, personnel, and of course supplies, materials, import duties, marketing, sales, receivables, and all manner of technical issues. We imported PV modules by the half container from Siemens Solar, Solarex, and BP Solar. Fortunately, by this time computers and the World Wide Web had replaced fax machines, and I could play my role from Washington more effectively, interacting in nearly real time with Colombo and Bangalore.

In order for this business to work, people had to be able to "access credit." But first they needed the credit to access. Rural banks in Sri Lanka continued to refuse to take risks on rural people without sufficient collateral. In contrast to banks in India, they would not take the SHSs our customers purchased as security, even if we offered to send out solar "repo men" to settle unpaid loans. So we turned to SEEDS, which by now had financed nearly all the solar lighting systems sold on the island to date through the SELF program. The legions of World Bank credit "experts" independently came to their own conclusion that only Sarvodaya had the "capacity" to become one of the bank's "participating credit institutions." All our early work had paid off, and the first World Bank solar rural-electrification project got under way. By 2000, Sarvodaya had financed six thousand SHSs. More competitors got into the game, and business boomed. In 2001 alone, Sarvodaya's financial institution had facilitated the sale of over ten thousand SHSs.

But just when success seemed to be at hand, the problems started. Everything moves very slowly in the developing world—maybe that's why it's still developing. Time is different there, palpably so. As the Sri Lankan supposedly said to the Mexican, upon learning the meaning of *mañana* (which famously expresses that you can always do tomorrow what you didn't do today), "Sir, we have no word in Sinhalese to convey such alacrity."

Tomorrow means next month on this languorous, steaming isle; a month means a year. What we in the West are accustomed to doing in hours takes many, many days in these unhurried lands. Time is culture, and they have plenty of both.

This meant that getting paid by the World Bank's ESD Project, the funds of which were managed by the government, would be done on Sri Lankan time, thank you. Not World Bank time, not American time, not SELCO time, not according to standard business time. Credit only works if the lender pays the supplier for the goods purchased by the borrower. Of course, you must allow time for a loan to be processed, especially one that originates in a rural area on paper, but thirty days should be sufficient. In the West, ten or fifteen days is normal for credit transactions on paper. Merchandisers and retailers get paid by credit card companies almost instantly. The entire Western economic system, which has built the mightiest economies ever known, is based entirely on the ease and efficiency of credit. America borrowed its way to prosperity (until we had our comeuppance in 2008!). Not so the two-thirds world.

Since most World Bank economists have never run a business, economic theory must suffice. And so the legions of bank economists who streamed in and out of Sri Lanka during those years to prepare the solar-finance program managed, in the end, to come up with a problematic project that was deformed at birth. Yes, it was funded, and yes, $30 million was now available for long-dreamt-of credit for solar loans. But then intractable local bureaucrats at the government "apex lending institutions" were put in charge, not private commercial bankers, and their intransigence made them the envy of Indian officials. Moreover, unbeknownst to all the entrepreneurs taking huge risks at their new solar enterprises, the solar project had been designed to eliminate all risk to the World Bank. Risk was forced downward, landing exclusively in the laps of RESCO, Suntec, and the other small solar companies the program had encouraged to come into being.

Nonetheless, the national solar program took off, and soon enough we were signing up farmers as fast as our fleet of motorbikes could carry our salesmen down every country road and up every jungle track. We would bring along the local credit officer from the

SEEDS program. The salesman would pitch our solar lighting system, and the farmer would apply for credit over tea in his house, and an SHS sale would be "booked" in a subsequent visit, followed by installation a week or two after. We had our sale, the customer was thrilled to have solarelectric power and light, and Sarvodaya had the loan papers, which now had to make their way through an impenetrable maze of steps that had been concocted by the visiting experts to avoid risk. And by the end of the steps, what was a World Bank loan to the government of Sri Lanka at rates of 1.5 percent became a 16 percent loan to our customers after each lending agency along the way took its quotient of "points" for risk.

Soon I learned to my chagrin that our Sri Lankan solar company wasn't getting paid in a remotely timely manner by the World Bank's illustrious ESD Program, now touted by bank officials at development conferences around the world. Receivables were running to 180 days. It's impossible to operate a sustainable and growing business this way.

Our managers in the field were often intimidated by the World Bank consultants, who continually parachuted in to supervise and "monitor" the project, and by the arrogant local project staff of the bank, who held the future of these companies in their hands. It was not the happy marriage of public and private interests for the common good that we had idealized might be possible. But we nevertheless struggled on year after year.

Compounding our troubles, it became clear that PW was not the world's best manager, and he was soon hard-pressed by these cash-flow and credit concerns and was putting our entire investment at risk. John D. Rockefeller once said, "You can build a friendship on business, but you cannot build a business on friendship." How right he was.

So in 2000, SELCO purchased RESCO's minority shares, and we renamed the company SELCO-Sri Lanka, changed all the signs and logos, and found a new manager, relieving the beleaguered PW to retire to academia. The company would not have existed without his tireless promotion of his dream, but it was time for him to go. I was the last to see it, but when I finally did, I took immediate action, as one must in business. He took the board's decision graciously, and we remained friends.

We hired a young, convivial manager, Susantha Pinto, who at the time was running one of Sri Lanka's largest tea plantations with fourteen hundred employees. We had electrified some of his "line houses," company dwellings for the Tamil tea pickers, with the tea company guaranteeing the loans, and Susantha had quickly noticed that solar energy was an interesting idea and more fun than growing tea! Before he was thirty he had reached the top level in the tea business: superintendent. He had no formal education but had done agricultural studies in Germany and Thailand. He wanted a change.

A bear of a man, Susantha occasionally picked up recalcitrant tea workers with one hand and dropped them, not in anger but to make a point: "I'm the boss." He oozed power and charisma and he was unflappable under pressure; nothing could rattle him. He was also capable of getting six jobs done at once. I had never met anyone in business anywhere with as much energy and drive. No one needed to tell him what to do, and so I didn't. I asked Harish only to give him advice. We announced the creation of SELCO-Sri Lanka at a well-attended press conference at the Trans Asia Hotel on May 23, 2001. Susantha quickly ramped up sales to several hundred SHSs a month.

Susantha was always very formal, insisting on calling me Mr. Williams. I never could break him of the habit. He also dressed formally, with a crisp suit and tie, despite the withering humidity of Sri Lanka. His employees also wore ties, even at the service centers, except for the technicians, whom he put in bright SELCO uniforms with embroidered logos on their shirts and hats (women, of course, as in India, wore saris).

I was always amazed at Sri Lankans' formal business attire, even in our small solar business, and I had to do as the Romans and wear a tie myself everywhere. I loved to joke with the casually dressed managers of SELCO-India that if they didn't shape up, SELCO would order a new dress code based on Sri Lanka. South Indians *never* wear ties. Funny, the cultural differences of two subcontinent nations and ethnic cousins separated by only a few miles of water. (A further note on apparel: Indians, Sri Lankans, and Vietnamese men and women always took pride in their appearance, always looked neat, and were always freshly pressed and clean, putting Americans and European travelers to shame for their slovenly dress, which has somehow become standard today. Rule one: no one wears shorts in India or Sri Lanka, but everyone is too polite to notice when the foreigners do.)

After we'd capitalized SELCO-Sri Lanka sufficiently, we inaugurated a new three-story headquarters building in suburban Colombo. Arthur C. Clarke, who had been knighted by Prince Charles a few months earlier during his visit to the island, accepted an invitation to speak at our event, but security for a simultaneous visit to Colombo by Chinese premier Jiang Zemin closed down all the roads in the capital, and he couldn't get there. Afterward, we trooped over to his house for an audience with the great man, whose Arthur C. Clarke Centre at the University of Moratuwa had been researching PV for years.

In the early 1990s, Arthur had agreed to narrate the opening of a video, *The Eternal Flame,* about our early projects in Sri Lanka. Willie Blake (not your usual tongue twister of a Sri Lankan name), Lalith's uncle, who earned his film stripes as a second-unit cameraman for David Lean's *Bridge on the River Kwai* (filmed in Ceylon), shot the video for us. In his introduction, Arthur says, "For the last few centuries since the industrial revolution, we've been living on our capital, on the energy stored up in coal and oil from sunlight hundreds of millions of years ago. We've been eating up these reserves at a colossal rate. In the near future, maybe only a few decades, the oil and gas and coal will all be gone. That means we'll have to go back to the original source, the sun."

Arthur later told *The New York Times,* "There are so many things coming to a head simultaneously. The population. The environment. The energy crunch. I feel rather depressed. . . . I think we have a 51 percent chance of survival. I would say the next decade is perhaps one of the most crucial in human history." Well, that decade is long past, and sadly we've made no real progress combating climate change or reducing population growth significantly. But the late Sir Arthur, whose science fiction books (and nonfiction writings [see *Greetings, Carbon Based Bipeds*]) foresaw great things for the world, would be pleased at the explosive growth of solar energy the world over.

Solar power gave rural people hope. As I've already mentioned, the greatest joy in this business has been visiting the homes of local farmers to witness their joy at finally having electricity. Over the decade of the 1990s I traveled thousands of miles on Sri Lanka's back roads, and at every house we visited I'd ask the residents how they liked their SHS. My colleagues would interpret for me. At one well-crafted stucco and tile house deep in the jungle, Mr. Sugathapala told me, "We were fed up with lighting kerosene lamps and carrying the battery to charge. Today, we have good lights to study by. Since there is no soot, the walls are nice as well. We are very pleased that SELCO was able to provide new light for our home."

A neighboring woman, Pushpa Samarawickrama, told me, "Solar power has changed everything for us. My children can now study late into the night, while there are no worries about bottle-lamp accidents. Most of the villagers are involved in self-employment projects such as farming, livestock, and sewing. Earlier, we could not work after six P.M. as it was dark. But now we can continue our work till late. What is more, this also drives away the wild elephants who used to roam at night. We have sunlight whether it is day or night!"

One farmer showed me his four bare rooms, two with sleeping mats, one for the kitchen, and the fourth for living and eating. There was no furniture. But there were light switches embedded in the walls, and he proudly turned on the lights, room by room, and said, "The last thing a farmer needs after coming home from the fields at the end of a long day is to fumble around in the dark trying to light a kerosene lamp. Now, we just switch on the lights, like in the city!"

This was our company's bottom line, right here; these were our real profits, and the profit to the community. It was especially rewarding when whole communities elected to go solar. This meant telling the local politicians, who had promised electricity but not delivered it, to not bother them anymore about grid power. Usually these efforts, in Sri Lanka, started with the village temple—specifically with the local priest, who in Theravada Buddhism is as much a social worker as a spiritual leader. Buddhism considered the sun a deity, and Buddha had said, "We give salutations to him, who is the sun, from where all golden rays emanate to light up this living world." We needed to get the Lord Buddha involved in our marketing program; he understood!

One priest, Venerable G. Dharmaratana Thero, said, "Before SELCO's arrival in the village, everyone used kerosene lamps and was facing many hardships. The SELCO systems have been immensely beneficial to carry out the religious and social activities of our temple." Evening activities in the temple courtyard improved markedly when the solar lights drove out the venomous snakes that formerly lurked there. I always got a kick out of how delighted the saffron-robed priests were when they could watch TV, which the temple committee usually bought for them once they got solar power. And after the community temple got lights, every villager wanted them, and SEEDS loan officers could not sign up people fast enough.

The young technicians we trained were quick to learn, and they should get the largest share of credit for making solar electrification possible. Lalith and PW personally trained dozens of technicians, many of whom became trainers themselves. Training courses, refresher retreats, and technical seminars were always under way somewhere to meet the need for skilled workers who could install SHSs in rural homes.

One of these was Chandral Chandrasena, a Sarvodaya Rural Technical Services employee, who quickly took to the technology. Chandral had an aura of distinguished contentment and a deliberate demeanor that focused him intently on his work of "right livelihood." He drove his Sarvodaya van over the roughest roads, to the remotest houses, and in the midday heat he climbed through the rafters, stringing wires, and onto thatched and tiled roofs to mount solar modules. The work was both mental and physical, requiring the skills of an electrician and the balance of a trapeze artist. The hours were long, six days a week, and month after month Chandral and his crew installed two SHSs a day for Sarvodaya. After each installation was complete, he patiently trained the appreciative family in its use. Chandral never complained or made a fuss. Since he was employed by Sarvodaya, his pay was minimal, but that didn't matter. He told me that he was doing "God's work" and that it brought him deep satisfaction. "Installing solar systems is a labor of love," he told me one day. "I like to share the joy and excitement when I see the village children experience electric light for the first time."

I always felt humbled by such examples, and amazed by people with this much dedication and principle. All the high-tech solar modules, all the sophisticated electronics, all the detailed financial planning and well-crafted business models meant nothing without a Chandral to install the apparatus in a rural house and show the family how to operate it. Where did these people come from? Sri Lankan, Indian, and Vietnamese culture fortunately produced many of them, many Chandrals, at least in places not yet contaminated by greed, consumerism, apathy, despair, and the suffocating demands of finance capitalism and crushing debt. Our solar systems brought hope to installers and customers alike. As Dr. Ari had written regarding Sarvodaya's self-help community projects, "We make the road and the road makes us." The same could be said for solar. Here was something worth doing, and worth doing well, and I always marveled at how well these guys—and many women—could do it.

What I learned in business in Sri Lanka, as we sought to explore "new economic paradigms" and socially responsible business models, is that you get the best results when people are motivated not by money, but by something intangible, such as a love of service or a passion for the cause. This, which I call the third way, was best described by the British business writer John Kay: "Individuals who are most successful at making money are not those who are most interested in making money." He cited numerous corporate case studies to prove it. I thought of the late Mohan Singh Oberoi, founder of the luxury Oberoi Hotel Group in India, who said: "You think of money and you cannot do the right thing. But money will always come if you do the right thing. So the effort should be to do the right thing."

This was SELCO's credo in Sri Lanka, India, and Vietnam. But some of our shareholders and their financial managers could not understand this. I was to learn the hard way that "the rights of capital" and "doing the right thing" are not always reconcilable or compatible, at least not in the eyes of many young investment analysts and portfolio managers, who learned everything there is to know in the world from business school. We knew we had to be economically profitable—it went without saying, and I loudly preached the priority of profitability over social service whenever I could. Thank God we mostly had patient capital, for now.

At the dawn of the new century SELCO-Sri Lanka had one hundred eighty employees and was selling and installing three hundred or more SHSs per month, sometimes hitting five hundred. Given that each sale was equivalent to an American family buying a car, this was a substantial business for rural Sri Lanka. The company had ten solar service centers around the country and over one hundred dealer-agents selling SELCO SHSs. Susantha was a celebrity, appearing on TV and in the newspapers and business magazines. However, the company continued to be plagued by its long-term receivables and operating debt.

Being the pioneer doesn't always guarantee success, for now we also had imitators and competitors. Shell Solar Lanka Ltd. was under the able leadership of Damian Miller, a tall American with movie-star looks who also ran the aforementioned Shell Solar India. Damian had learned everything he knew about solar from SELF and from studying SELCO-India as a MacArthur Fellow. His Oxford Ph.D. was about our India operations! He ran a tight ship, and Shell quickly took half our market share. Lal Fernando told me, "Shell selling solar is like an arms dealer running a refugee camp." Well, maybe. I never knew what Shell's game was, except that it spent tens of millions of dollars advertising its commitment to solar energy in double-page spreads in *Time, Newsweek, National Geographic*, and just about every other national and international magazine.

I've already mentioned their global TV ads featuring Damian Miller. Damian turned out to be a crackerjack executive, overseeing complex and far-flung operations in Asia and Africa from his base in Singapore and giving SELCO a run for its money. Since Shell was just getting into the renewables game in 1998, before he became our prime competitor in India and Sri Lanka, Damian thought it would be a good idea if Shell just bought us. So I was invited to the top-floor boardroom of the Shell headquarters building in London, overlooking the Houses of Parliament and the Thames. Sitting around an enormous boardroom table, two Shell executives, in their words, told me, "we can make you an offer you can't refuse."

For the first time, a huge multinational had put its brand name directly on solar power technology and products, and now it wanted to brand "rural solar" in the developing world. We'd already branded it, with SELCO's bright-red sunburst logo, and Shell wanted a shortcut to these hard-to-reach markets that we'd spent so many years cultivating. It wanted to avoid our painful years of trailblazing and get right to some of those two billion without power, *now*.

"We could call the merged company 'Shellco,'" they joked.

SELCO had not yet even begun to create value, so at this stage it wasn't worth very much. Which of course is why Shell didn't offer very much. In fact, the offer was so ridiculous that SELCO's board broke up laughing when I presented it to them.

Years later, in 2002, Damian would request my support for the Million Homes Initiative he had proposed to the Global Environment Facility. His idea was that G8 countries would make an additional combined donation of $150 million to the GEF to subsidize SHSs at the flat rate of $150 each. Having witnessed what worked and what didn't work in the solar energy world, he had proposed a fair and straightforward disbursal mechanism that would allow commercial entities to install one million SHSs in five years. He wasn't suggesting the World Bank kick in any money, only that donors to the GEF put up an additional $150 million targeted exclusively for this program—almost exactly what the United States at the time was spending *per day* in Iraq.

It was an entirely workable idea, and the World Bank president, James Wolfensohn, liked the idea. I was invited to several meetings in Washington in the spring of 2003 to explore this breakthrough solar-electrification proposal with officials at the GEF and the World

Bank. It didn't take more than two months for bank bureaucrats to kill Damian's Million Homes Initiative, and it remains deader than roadkill.

Meanwhile, Shell Solar Sri Lanka was experiencing the same cash-flow problems that bedeviled us, caused by the bank's Energy Services Delivery Project's glacial pace of refinancing our customers.

One day, our auditors, Ernst & Young, reported that they were going to "write down" about $150,000's worth of receivables owed us by the World Bank as bad debt. "The World Bank as bad debtor!" I howled back in Washington, where I promptly called the bank.

"How can we run a company like this?" I asked them.

The program managers made all kinds of excuses and passed the buck when I explained that the bucks weren't getting passed to us. This is all too tedious to recount further, but the good news was that SELCO-Sri Lanka operated for a few more years, even profitably during some quarters, continuing to deliver rural electricity on a lush, mountainous, tropical island until its luck ran out and our parent company's board lost interest in further capitalization. Susantha Pinto was signing up people faster than he could serve them, faster than the Sarvodaya and World Bank loan programs could process them, and cash flow continued to be the paramount issue for the company. You have to be paid for the goods sold to be able to order more goods. SELCO-Sri Lanka's sin was to leverage itself beyond its ability to meet obligations to vendors, while focusing on the ever-expanding market, booking sales and reporting income in advance of collecting the funds, and digging an ever-deeper hole while sales continued to grow. In business terms it's called "selling yourself to death." The "company that could" folded.

A dozen locally owned competitors somehow kept going. Eventually, Shell Solar Lanka also closed down when its oil parent pulled the plug. Sri Lanka continued to be a crucible for the rural renewable energy business, and yet its small market of fewer than a million families was of negligible commercial interest compared with that of the giant to the north, India, with over 100 million families (at least 400 million people) who needed reliable electricity that only solar power could provide.

In the years that followed, the Ceylon Electricity Board, over the objections of local environmentalists, decided to build two coal-fired electric generating plants to provide power for a 100 percent subsidized grid extension into most rural areas; hookups were free. (Remember the politicians' promise: "Electricity for all by the year 2000!"?) If solar power had been free or heavily subsidized, we would have covered the entire island, but as solar developers worldwide are learning, the power companies and the coal lobby often have the last word, and they buy the government they need to do their bidding. Just as in the United States. And how about this: Sri Lanka *has no coal,* and must import it all from Indonesia! The sun, unfortunately, has no lobby as powerful as the worldwide coal lobby, although Greenpeace's 2012 global Quit Coal campaign was beginning to have some success as more and more nations shut down coal-fired generating plants. Yet in Sri Lanka, they were now building them!

Today, only a few solar companies remain, some serving the former rebel territories in the north recovering from the twenty-six-year war, some gearing up for large grid-tied PV systems that will certainly be able to compete with imported coal and with oil for diesel generators. Twenty percent of the rural population is still without electricity, so opportunities still exist for the remaining entrepreneurs, but this market is hampered by lack of credit now that the donor programs are over.

But there is also good news. Anil Cabraal, now retired from his career at the bank to his native Sri Lanka, where he still consults on rural energy worldwide, sent me the final report by the United Nations Development Program of the ESD (later called RERED, or Renewable Energy for Rural Economic Development project). It showed that by 2011, when the World Bank solar-investment programs were concluded, 127,069 Sri Lanka homes had been able to purchase solar home systems, all financed through Sarvodaya and several local "credit institutions." (The same programs also included minihydro for another five thousand households and village-based hydroelectric systems that provided an additional 120 megawatts to the national grid.) Now, more than one million people are no longer in the dark. Studies showed that the average family using solar PV has reduced its monthly kerosene consumption from eleven liters to nearly zero. This reduction improved health, reduced greenhouse-gas emissions, and saved money. And people could now charge their mobile phones, and they no longer needed to purchase batteries for their radios or go to town and pay to have a car battery charged to run their TVs.

Fifteen years earlier, Anil and I, as well as the Suntec guys, plus Dr. Ari, Lal, and PW—and, yes, all the well-intentioned World Bank staffers, technocrats, and businessmen—had, together, really started something on this serendipitous island, and for that we could all be proud.

Best of all, Sri Lanka was finally at peace.

8

"War Is Over If You Want It": 1968–2001

War is never really over—in spite of what John Lennon and Yoko Ono proclaimed from hundreds of billboards around the world in 1969—but the Vietnam War finally ended for me in 1993, when I first returned to the land of our fourteen-year "conflict."

Memories flooded back when I visited the Museum of American Aggression in Ho Chi Minh City, as Saigon was now called, and re-visited Chivral, the downtown coffeehouse where GIs and journalists used to buy marijuana cigarettes, perfectly rolled and wrapped in cellophane-covered Salem packs—cartons of them. Across from Chivral, the old Continental Hotel had become a four-star property with a fine Italian restaurant on the veranda, and, opposite, the old Caravelle was being turned into a spectacular luxury hotel and condominium tower. The rooftop Caravelle Bar still offered the same panoramic views of the former Paris of the Orient: the broad boulevards, the Saigon River, the brick cathedral, the opera house below. From here, as a young freelance correspondent, I had watched the phosphorous flares illuminating the city's perimeter at night and felt the low rumble of B-52s unloading vast quantities of bombs on the Mekong Delta. Now, there were tall office towers sprouting up, including the Citibank building across Le Loi Street, where we opened our SELF and later SELCO accounts.

In 1968, with 535,000 Americans in-country, the war we were destined to lose was at its peak. I arrived not long after the famed Tet Offensive and spent part of 1969 flying along on bombing missions with the Colorado Air National Guard, accompanying draftee "grunts" on jungle combat patrols, visiting remote firebases, and watching the Twenty-fifth Infantry Division completely destroy, for no good reason, a group of pastoral hamlets, home to several thousand peasant farmers, west of Cu Chi. I visited my friend Ollie Davidson, who would later join the board of SELF, at his compound in Trang Bang, where he was in charge of "civil operations" for USAID. There I was, cajoled into joining a night patrol of the local anti-Vietcong militia and given a captured AK-47 to carry, along with my camera, just in case.

Back in Saigon at one of Time-Life photographer Tim Page's legendary substance-abusing parties at his Tu Do Street apartment, Sean Flynn (son of actor Errol Flynn) asked me to ride with him to Cambodia on his motorbike to see the war from that side. I nearly went, but thought better of it. The war-loving adventurer-photographer later made the trip with Dana Stone, a CBS reporter, and the pair were never seen alive again.

But all that was so, so long ago, and I was finally getting back to Vietnam nearly twenty-five years after those misspent days of rage and glory.

Vietnamese Women's Union Conference on solar rural electrification in Hanoi, 1999; author, second from right; Pham Hanh Sam and Ho Chi Minh, far left. *(Author)*

SELF was in Vietnam, as I explained in chapter 3, because a third of the population, six million families, most of them rural, had no electricity, and the Vietnam Women's Union (VWU) had agreed to do a joint project with SELF, and later SELCO, to electrify their rural members with solar power. Madame My Hoa, president of the VWU, introduced me to Madame Phuong, a powerhouse who managed huge social-development projects for the VWU like a Prussian general. She was assisted by a shy younger woman, Madame Pham Hanh Sam, who eventually took over the projects in Vietnam when Madame Phuong retired. (Retirement is mandatory at fifty-five— life expectancy is much shorter in the two-thirds world.) While both women spoke some English and could read it, they were nonetheless always accompanied by capable young female interpreters who spoke English perfectly.

Madame My Hoa arranged for me to go with Madame Sam, local VWU officials, and an interpreter to visit a hamlet near Ho Chi Minh's birthplace in Nghe Anh Province, about two hundred kilometers south of Hanoi. The provincial capital, Vinh, had been flattened by U.S. bombing and completely rebuilt by the East Germans, rendering it the ugliest city in Vietnam. We negotiated dirt tracks in our Chinese army jeep, traveling thirty-seven kilometers west of Vinh, past the last electric power lines, and on through the low jungle scrub to the self-contained hamlet of 1,280 households. The People's Committee of Thinh Thanh had set out orange sodas and cookies in the community center, which I had promised SELF would electrify to showcase solar PV in the province. I asked if they had ever met an American before.

"We have not seen an American since one came down from the sky with a parachute during the American War," said Mr. Dung, the thin, angular, People's Committee chairman. "We took him to the authorities." The chairman seemed not to wish to open this embarrassing chapter of our joint history, however, and I let the matter drop and smiled. "Let bygones be bygones," he said.

Turning to the matter at hand—electricity—the district People's Committee chairman told me that Thinh Thanh would not get electricity for ten years. The commune officials thought maybe sooner.

He informed everyone that it would cost the commune the equivalent of $1.2 million to bring it here. There were no more subsidies since *doi moi* ("open reform") began in 1986.

Mr. Dung said, "We will take care of the solar project. People will pay. It is not like before when things were free. We want to bring the light to improve people's lives."

For less than half a million dollars we could have electrified everyone with solar, but we didn't have that kind of money. With the first grant funds I had previously raised from the Rockefeller Brothers Fund we could electrify thirty households, plus the community center (for which the hamlet would have to pay the women's union). Thirty families had already signed up. The local VWU official was very excited about the program and agreed to collect the monthly installment payments.

The average annual income here was $400 to $600 per household. Our smallest SHS cost us about $330, which SELF passed along without markup, but that was still a lot of money, even on a four-year low-interest credit term. Yet the people eagerly paid their deposits to get on the list. Besides growing rice and vegetables, they earned income by contract sewing. But there was no light to sew by in the evenings, when most of this extra work was done. Our lights, we later learned, had a big impact on the community.

Down the spine of Vietnam, from the huge Hoa Binh hydropower project in the north all the way to Ho Chi Minh City, a huge 500-kilovolt transmission line stretched across 150-foot towers spaced thousands of feet apart. With sufficient hydro resources in the north and growing power demands in the booming south, this was an internal Vietnamese solution to a Vietnamese problem: they had proudly built the line themselves. The amazing thing about it, however, was that the huge high-voltage line rose and drooped, from tower to tower, across a thousand miles of inhabited land. Tens of thousands of people lived within sight of, even directly underneath, these soaring electric cables yet had no access to electricity. However, everyone to whom I pointed this out was smart enough to know that it was too costly to step down 500 kilovolts for residential service, and that if communities along the way tapped this power, there would be none left for Ho Chi Minh City. Our jeep passed back and forth under this power line as we drove to and from Vinh and other communes.

As Sam and Phuong and I rode back to Vinh with the members of the local women's union, it seemed that bygones were not gone. A woman in the community we had just visited told me that she had just missed direct hits by U.S. bombers three times and that sixty-four people in her village had been killed. American authorities said the United States was just targeting the nearby bridge. As a journalist many years earlier, I had interviewed fighter-pilot POWs who had been assigned to hit that particular bridge north of Vinh she was talking about, which we had driven over. The pilots said they were never able to knock it out.

Sam lightened the moment with her own observation. "When we were children, it was very exciting for us when the American planes came over—we saw many, many. We liked to hear the 'boom, boom.'"

As I flew down to Ho Chi Minh City, I saw more reminders of the war: the row upon row of hardened concrete revetments at Tan Son Nhat Airport that once housed our Phantom fighter jets. They're still there, good as new. No one could knock them down; they had been built by Kellogg, Brown & Root, now part of Halliburton. Ho Chi Minh City mostly looked the same, with rivers of motorbikes flowing up and down the tree-lined avenues. Many were driven by women with flowing *ao dais,* the formal pant-dress, and many perilously carried whole families, stacks of produce, or cartons of appliances through the nightmare traffic.

We were to meet Madame My Hoa in the delta at the provincial river town of Tra Vinh. We boarded boats, hers a naval patrol craft and ours a small, wooden inboard, and headed through the Mekong's maze of islands and canals to reach the hamlet of Phu Dong. The women on board, all members of the national, regional, district, and local VWU, plus Sam and Phuong, broke out fruit and soft drinks on the deck and had a party as we navigated the kind of narrow waterways that had been memorialized in the film *Apocalypse Now.*

Madame My Hoa, in fact, had led a local Vietcong unit in these low-lying islands and mangrove swamps; this was her home, although she was now based in Hanoi. She had been captured during the war, I learned—although she never mentioned it—and had spent many years in prison. She was welcomed at Phu Dong as a returning hero. Phu Dong had been selected as another site for our solar project because it was a "revolutionary area." This meant, unfortunately, that the local "war hero" families expected to get their solar systems for free, as an entitlement for their service in the American War.

Hundreds of village children crowded around, staring wide-eyed at the foreign visitor. Madame My Hoa said, "See all the people that have come to have a look at you!" The local VWU representatives told her that they had already signed up one hundred families for the SELF-VWU solar program and had collected the initial deposits. "They are saving for the light," the local VWU official told me. "Many people are suffering from living and dying in darkness."

I certainly never imagined, when I flew over the green expanse of the delta in U.S. helicopter gunships in 1969, that one day I'd be down here in this dense, stiflingly hot undergrowth selling solar-electric systems to peasant farmers and former fighters for the Vietcong.

Phu Dong and its neighboring hamlet, Phu Tan, were home to some three thousand families who grew rice, fished, and raised shrimp in large excavated shrimp ponds. Because they lived on delta islands separated from the mainland by miles of water—the Mekong's fingers are amazingly wide here—they had no electricity and no prospects of getting any. Perfect for solar.

The next time I returned to Phu Dong and Phu Tan, a large solar array of huge, glass, amorphous-silicon modules stood on a cement-anchored frame in front of the community center, VWU offices, and clinic. The 300-Wp experimental thin-film modules had been donated to SELF, and I'd arranged for them to be shipped to Ho Chi Minh City with one of our container loads (this is a common module size today, but wasn't then). They produced more power than the center knew what to do with. They had transported them by boat and installed them here, along with a hundred 20-Wp SHSs installed in households scattered through this very remote community. These distant hamlets could boast of being the genesis of solar rural electrification in Vietnam, and the women's union was very proud of its ability to undertake a technical project of this kind.

"I was very worried about this new technology, and worried that people couldn't pay back," the local VWU president told me. "But the solar works, we overcame our difficulties with it, and now the light contributes to the people's material life, their literacy, and gives opportunity for income generation and expanded knowledge." I couldn't have expressed it better myself. "Seeing is believing," she added, "as we say in Vietnam."

To assist the women, I sent an experienced PV technician to Vietnam: Marlene Brown, a former solar-system installer from New Mexico who was then working as a technical expert at Sandia National Laboratories (which helped finance some PV systems in Vietnam long before the United States had reestablished diplomatic relations). Marlene, who also worked for SELF in the Solomon Islands, had

Project manager and trainer, Marlene Brown, raising a thin-film solar panel in the Mekong Delta, Vietnam. *(Marlene Brown)*

more enthusiasm than just about anyone I had ever met, and she also possessed the technical skills required to train, with the help of interpreters and Madame Sam, dozens of local women to install and maintain SHSs. She was fearless, tireless, funny, and always the diplomat who could overcome any confused situation with humor and understanding. Marlene and her all-woman crews could install ten SHSs a day, nearly one hundred a week, in conditions that would drive any normal electrical contractor from the West to despair. Marlene patiently taught every family how to use their SHS, and she photographed every solar house SELF electrified. She spent many months in Vietnam during several visits and became close friends with her contemporaries in the VWU, who adored her.

The only thing the women couldn't—or wouldn't—do was climb onto roofs to mount the solar modules. "Do you mind if we ask the boys to do it?" Sam asked me shyly one day. "They have offered to help us. Women don't climb on roofs." But they did manual work, such as carrying heavy batteries to the boats for loading, then unloading them on the islands and carrying them on poles to the individual houses. Transporting hundreds of solar modules to these remote locations was equally burdensome for the women.

In Tra Vinh Province, farther south, we launched projects in two more Mekong island communes, Hoa Minh and Long Hoa, home of some six thousand rice growers, shrimp farmers, and fishermen. To visit these houses, we walked along narrow dikes separating the paddies, crossed single-log "monkey bridges" over the irrigation canals, and traveled muddy paths through the thick vegetation. Every once in a while a simple thatched house with a tall pole supporting a solar module would appear, and the family would welcome me inside to see their lights, their TV, and often their stereo, some of them equipped with a microphone to be used as a karaoke system, all powered by our PV systems. In one house, an enormous python rested in his cage with a duck sitting on top of him, unawares. The owner said, "When he gets hungry, the duck is lunch." He would eventually sell the python for its meat and earn funds to put toward his solar system.

The president of the Tra Vinh Women's Union, Madame Khanh,

never smiled or showed the same enthusiasm Sam and Phuong had for our project, and I learned why: her husband had been killed by U.S. forces, and I was the first American she'd ever met. It was a testament to the humanity of the Vietnamese, and their attitude of national forgiveness for Americans and our war, that they would even talk to their former enemy, an enemy who had come from ten thousand miles away to destroy their land and kill close to a million of them. (Their humanity stood in stark relief to the suicidal barbarity of the Islamists in the Middle East and Iraq and the maniacal hatred harbored by the 9/11 terrorists.)

SELF and the VWU launched six separate solar projects in Vietnam, two in the north and four in the south. One of the northern villages was near Hoa Binh and the country's largest hydroelectric dam, but the people had no power. They earned money to pay for their SHSs by raising puppies for the dog restaurants in Hanoi. I was glad this was one community I never visited; I only saw the photos. (I didn't tell my wife, the executive vice president of the Humane Society of the United States at the time.) We installed several hundred systems in Tien Giang Province, also in the delta, where I suffered through embarrassing drunken noontime feasts with the local People's Committee and Communist Party chiefs. Sam and Phuong were appalled, but there was little they could do when a local party functionary decided a visit by a foreign guest was an occasion to drink himself and his buddies under the table on the government's tab. The food, however, was terrific, except for the duck eggs with fully formed, unhatched ducklings inside—very good with the fermented fish sauce called *nuoc mam,* I was told.

In Hanoi we invited the government bureaucrats from the ministries of industry, finance, planning and investment, science and technology, agriculture and rural development and representatives from Electricity of Vietnam, the World Bank, the UNDP, and the Rockefeller Brothers Fund to an all-day "national solar-electrification seminar" at the top-floor conference center of the VWU's headquarters building. Women came from all over Vietnam, from all levels of the VWU, dressed in their finest *ao dais*; they were there to learn about solar or to provide testimonials on how solar electricity had changed lives in their homes and communities. We were seated on a dais beneath a large plaster bust of Ho Chi Minh; simultaneous translation through headsets allowed speakers to use Vietnamese or English. All day long we heard how electric light had transformed lives, how the local women had struggled to implement the financial and technical aspects of the project, and how proud they were of their ultimate success, electrifying some one thousand families with no government support.

We shipped containers of U.S.-made solar modules and batteries to Vietnam, duty free, and Madame Sam and her patient staffers managed to get the shipments through the corrupt customs (eight customs officers were executed for corruption while I was there) and delivered to their storage sheds on the grounds of their southern headquarters in Ho Chi Minh City. The VWU's offices there were located in the former villa of General William Westmoreland, who commanded the U.S. war in the 1960s.

It was in General Westmoreland's old living room that I attended a private dinner with the woman in charge of the southern branch of the VWU, Madame Thang. In her mid-forties, she was stunningly attractive with a warm smile and a no-nonsense manner. She wore a colorful silk dress, not an *ao dai,* and shimmering jewelry. Women's union staff served us from the villa's kitchen as we talked about the solar project and the good work Marlene Brown was doing. As dinner progressed, we reminisced about the war years.

"I spent six years in prison," she offered, "from 1969 until liberation."

"For what?" I inquired.

"I was a member of the National Liberation Front," she said, smiling.

She hesitated to say more, but I inquired. My own antiwar credentials helped me here—we both knew some of the same people in the U.S. antiwar movement. I told her a close friend had started Vietnam Veterans Against the War.

So she put down her chopsticks and said, "I tried to assassinate one of the top-ranking ARVN [Army of the Republic of Vietnam] generals . . . in his bed."

"Oh," I murmured.

"I had pretended that I would be his mistress so I had access to the house, and one night I went to his bedroom, lifted the mosquito net, and pointed a big pistol at his head . . . but I couldn't do it. Maybe I could have, but before I could decide, he woke up, grabbed me, and I was arrested."

As a lovely young woman in 1969, she was a celebrity at her trial in Saigon, where a judge sentenced her to life in prison. When the verdict was read, she smiled what became a very famous smile, photographed by the AP and sent around the world by newswire. The judge asked her, "Why are you smiling? I just sentenced you to life in prison. That is the smile of a crazy person."

"No," the young Madame Thang answered, "it is the smile of victory."

She was let out of prison when the National Liberation Front and North Vietnamese Army overran Saigon in April 1975. The "smile of victory" photo still hangs in the Vietnam Women's Museum, which chronicles the role of women in Vietnam since the Trung Sisters drove out the Chinese invaders in AD 40.

Shawn Luong, a senior engineer with the Los Angeles Department of Water and Power, contacted me one day in Washington after reading an article I wrote in *Solar Today,* a trade magazine.

"This is the right technology for Vietnam," he said. "I'm familiar with PV. I've installed many mountaintop communications towers in California for the Department of Water and Power [LADWP], powered by solar. It would be good to use PV in Vietnam, where rural people have no electricity."

I said we already were.

Shawn flew to D.C. and offered to help. He also wanted to set up a joint venture with SELF in Vietnam, and I said we were considering forming a company, as we had already done in India and were about to do in Sri Lanka. Shawn (not a regular Vietnamese name, as one might suspect) and his parents escaped from Vietnam in 1979, four years after "liberation." His father owned lumber mills, textile factories, and farms near Saigon, which were confiscated by the new government. Shawn, who spoke French as well as fluent English, had attended the Lycée Madame Curie; he was twenty-one when they set off on a small boat, survived pirates in the South China Sea, and washed up in Thailand, where they were taken to one of the big UN refugee camps set up for the boat people. Being a natural hustler, it wasn't long before he managed to get his family, his future wife, his brother, and himself to Los Angeles, arriving with a total of two hundred dollars in their pockets. The new Communist government had confiscated all their savings.

Shawn came from the Chinese community in Ho Chi Minh City, which had been there for several hundred years, blending in but retaining the language. He told me, "We Chinese are the Jews of the Orient." Chinese, Jewish, Vietnamese—I found Shawn immensely likable, honest, and hardworking, although extremely cocky (to use his own word). I figured I could not find a better partner with whom to launch SELCO-Vietnam. I certainly couldn't do it myself; I didn't speak the language. He had been running a trading operation on the side and had gone back in 1990 to meet the rulers in Hanoi who had driven him and his family out of the country fifteen years earlier.

With a graduate degree from the University of Southern California and a successful twelve-year career at LADWP, he was ready for something new. We agreed his trading company offices in Ho Chi Minh City would be the new SELCO-Vietnam headquarters, and he set about securing the first license for a foreign-owned solar company to do business in Vietnam. In December 1997 we received a nice blue-covered bound book containing our license to sell solar devices and equipment in Vietnam through SELCO-Vietnam Company Ltd.

Shawn quickly made friends with the VWU representatives, including Phuong and Sam. Together we approached the World Bank mission and the officials at the Ministry of Industry and at Electricity of Vietnam. We had Big Plans! With the biggest development organization in the country behind us, and connections in the politburo and a track record of hundreds of installed SHSs, why shouldn't we think big? Communist governments like big plans, usually in five-year increments, so we set out to develop an ambitious five-year solar-electrification plan with the support of the VWU. It would be the largest commercial solar project in the world. Or so we hoped.

It was in Vietnam that I was to learn most of the lessons I would need to manage in business, politics, economic development, corporate operations, human relations, and global finance. And I learned what cultural barriers were—barriers.

Shawn got to work, mounted a neon SELCO sign on his five-story glass office building on Nguyen Thi Minh Kai Street, electrified the SELCO office floors with solar PV to assure power for the lights and computers during power cuts, and we opened for business. We put Tran Thanh Danh, a self-taught electrical engineer and an energetic worker with a beatific face and infinite patience, in charge of operations.

The VWU identified remote villages where the local leaders verified there was little probability of electricity arriving and where people expressed interest in acquiring a solarelectric system. We expanded operations in areas where SELF had been—it was no longer operating in Vietnam now that SELCO had taken over the projects—and we opened solar-powered solar service centers. We moved into new provinces—Phuoc Vinh, Dak Lac, Ca Mau, Binh Thuan, and even Phu Quoc Island off the Cambodian coast. We opened service centers as far north as Ban Me Thuot in the central highlands. Shawn and Danh were spreading us pretty thin.

I transferred from Citibank Washington to Citibank Ho Chi Minh City a large chunk of money from the parent company to SELCO-Vietnam, the minimum capitalization required by our Vietnam business license. We then borrowed another large sum from the World Bank's division that makes loans to private companies. They were amenable to lending money to finance solar rural electrification in Vietnam, thanks to local staffers who truly understood development banking. A bank official had visited one of our service centers in Tra Vinh during a power outage, and the only light on the street—even his hotel was dark—emanated from the solar-electrified SELCO-Vietnam office.

To close the loan with the bank, we brought Phuong and Sam and their smart young interpreter to Washington via LA, where Shawn took them to Disneyland in his big black Mercedes and on to see his suburban split-level home on a beautifully landscaped hillside overlooking the Pomona Valley, where the women asked him pointedly, "Why are you leaving all this to come back to Vietnam to help us?"

I'd also wondered.

"My priest at our Buddhist Temple in Garden Grove told me I had to do it," Shawn said. "My karma required it." People had often asked me the same thing—"Why do you do this?"—but I never had such a clear answer. I wasn't doing it to get rich, that was for sure. I discovered the hard way that this rocky, risk-laden path was not the road to wealth. If wealth were what I sought, I could have found a far easier way to acquire it. So Shawn and I were somewhat on the same page in this regard.

In D.C. I arranged a formal lunch for the VWU delegation with a World Bank vice president at one of the executive dining rooms in the bank's gleaming new building on Pennsylvania Avenue. The bank thought the VWU should borrow the money, since the solar loans were going to be made to VWU members, but the women were smarter than that; they didn't want the liability. They would gladly take grant money and spend it wisely and honestly, but they rightly feared debt. So SELCO took the loan, along with all the risk and all the liability. Our company's future now depended on collecting installment loan payments from peasant farmers in Vietnam.

In 1999 we proposed a twelve-thousand-house solar project as the first phase in a plan over the coming decade to reach a million homes. The vice minister of industry, in charge of electricity for Vietnam, told me in a formal conference room at the ministry, with its plush chairs, lace doilies, lacquered coffee tables, and porcelain tea sets, that "we can never electrify these people," referring to the last two million families who would still remain in the dark twenty years hence. "We must use solar," he emphasized, pounding his chair arm. We gained the tacit support of Electricity of Vietnam, whose dour, skeptical engineers had come around to our view that solar PV provided the cheapest and fastest way to electrify communities remaining far beyond the grid.

Shawn set up a large warehouse on Ho Chi Minh City's outskirts, and we began manufacturing our own charge controllers and light fixtures and began assembling energy-efficient AC and DC bulbs in our workshop. Soon, half the hotels in the city were illuminated by SELCO-Vietnam's low-cost, high-quality compact fluorescent lights, and our DC bulbs brightened the homes of our growing base of rural customers. I was constantly amazed at the technical ability of SELCO-Vietnam's employees; they could fabricate just about anything, from PV-powered railroad signal lights to lamps with LED clusters, to solar streetlights and DC fans. LEDs, which we first used in Vietnam, would revolutionize the off-grid solar industry.

On the marketing side, an enormous SELCO billboard, illuminated by two hundred watts of solar power and our biggest compact fluorescent lights, towered over the central plaza at one of the Mekong River ferry crossings. It proclaimed in Vietnamese, "Solar Light Brings Joy to the Home!" With the VWU, we set up booths at agricultural trade fairs, where we displayed our solar product line to the amazed rural and small-town clientele. The VWU hostesses sported SELCO baseball caps; our technicians wore SELCO polo shirts.

I visited some of our customers in Binh Phuoc Province, out near the Cambodian border, just beyond the so-called Iron Triangle, where the United States had denuded over a thousand square kilometers of forest cover with toxic dioxins (Agent Orange) and removed the entire population to fortified "strategic hamlets" surrounded by barbed wire. With no people, it was presumed the Vietcong couldn't fight a "people's war." The people became refugees in their own land. This was the bright idea of one Herman Kahn of the right-wing Hudson Institute, which would later advise the U.S. government on Iraq.

Now, this fractured landscape, with its new growth, had been opened to recolonization, and families from the crowded Red River delta in the north had been offered land to farm in the Iron Triangle since the Americans had so kindly removed the trees. These were

SELCO's customers, growing black pepper on neat plots, nursing the land back to health.

In one two-story wooden house half covered in blooming bougainvillea, with mounds of peppercorns drying in the courtyard, Danh introduced me to a middle-aged peasant woman with a wide smile. She proudly showed me her house and switched the lights on and off in each room. She said she had more lights upstairs, where her son studied most of the night. I asked how the system was working. She had bought a 75-Wp system, our largest.

"Oh, fine," she said. "But when my son studies, sometimes we don't have enough power for our color television." She switched on the set, which seemed to be working, and I marveled that she could afford a color TV. (Our technicians rewired color TVs to operate on direct current.)

"What do you do then?" I asked. "Tell your son not to study anymore? Or stop watching TV?"

She promptly showed me her answer, opening a wooden cabinet on which the color TV stood. Inside was another television. "Oh, we

A pepper farmer in Vietnam shows off dual 12-volt TVs, one color and one black-and-white. She uses the black-and-white TV to save power when her son is studying by solar light. (*Author*)

watch our black-and-white TV. It uses less power." She had figured out what we called "power management." If you used solar power wisely, there was usually enough for all occasions during most of the night.

In other nearby houses, I found pepper farmers with excited children watching taped programs on their solar-powered VCR in black-and-white, since they couldn't afford a big enough PV system to run lights, a VCR, and a *color* TV. Most of the families had also bought a SELCO fan to move the humid air through the main room on a hot day. All through this revitalized war zone, one hundred kilometers northwest of Ho Chi Minh City, SELCO lights were keeping the dark tropic night at bay.

Vietnam is a long, skinny, coastal country, with its capital, Hanoi, about twelve hundred kilometers north of its commercial center, Ho Chi Minh City. Vietnam Airlines' air shuttle connected the two with Airbus 320s and Boeing 767s. (I had earlier flown on one of VA's last Russian Tupolevs before it crashed, as they all seem to do eventually.) Nothing happens without government approval at the highest levels, which can only take place in Hanoi, so Shawn found himself spending a great deal of time there. He wined and dined a friend in the prime minister's office, who promised to help us launch a huge national solar-electrification program. First, however, we had to sell the idea to the national assembly.

Shawn and the VWU worked together to put a SELCO exhibit right smack in the middle of the grounds of the national assembly during its annual meeting. It was a small, solar-powered, prefab house, with recessed solar lights and PV modules integrated into the roof. It was a sensation, visited by the media and hundreds of members of the assembly. No private company had ever been allowed to display its goods or products on the grounds of the national assembly. Shawn wore his best pinstriped suit, Madame Sam her best silk *ao dai*. Then, Nong Duc Manh, chairman of the assembly, came by for a visit, and Shawn and Sam explained to him how solar electricity worked. He was fascinated. Shortly thereafter, Mr. Manh was named secretary general of the Communist Party, the country's highest leadership post. No capitalist enterprise in Vietnam was complete without the blessing of the country's top Communist. So it goes.

Nonetheless, doing business in Vietnam was not easy. We learned the hard way. SELCO managed to install only 2,200 SHSs in its first few years, and the company found itself financing many of these customers itself, not good for the parent company's balance sheet. Despite the tireless efforts of the VWU to institute their solar credit program, and notwithstanding an alliance with the Vietnamese Bank for Agriculture and Development, microcredit, so successful in India and Sri Lanka, simply didn't work, and doesn't to this day. There is no history of, or experience with, credit of any kind in Vietnam, we were to learn later instead of sooner. Hopeful that the VWU, which vouched for the customers, would continue to collect the payments, we extended company credit, more and more credit to more and more families, scattered farther and farther away, all over southern Vietnam.

We prepared detailed loan documents in Vietnamese and English, which were signed by the farmer, the local VWU rep, and the local People's Committee, which is like having the sheriff cosign your auto loan. We planned to refinance these loans one way or another; they were scanned and e-mailed back to Washington, where one day I showed several hundred of them to the vice president and treasurer of the U.S. Overseas Private Investment Corporation (OPIC), who had come by the SELCO office in Chevy Chase.

The OPIC official told us, "These farmers are probably a better credit risk than some of the big power projects we finance." How right she was: our farmers continued paying their installments, while Enron stiffed OPIC and U.S. taxpayers for the $700 million it owed on its power plant in India. Unfortunately, SELCO's balance sheet didn't qualify us for an OPIC loan to refinance our farmers in Vietnam, so we were stuck carrying these receivables ourselves (few companies can finance the sale of their own products). Perhaps we needed Enron's accountants to fiddle with our balance sheet.

SELCO-Vietnam moved into new quarters at the beginning of the new century: a five-story building of our own on Tran Hung Dao Boulevard. The neon SELCO sunburst logo was visible from a mile away atop the building's roof. A clean, modern showroom graced the ground floor, open to the sidewalk, and sometimes people walked in and ordered a PV system for their urban home as a backup against annoying power outages.

Up in Hanoi, Shawn and I met with the U.S. ambassador, Pete Peterson, a former POW and ex-congressman. We told him what we were up to, and he offered all the support he could through the various U.S. agencies now working in-country. Pete, like Senator John McCain, had spent years in the "Hanoi Hilton," the old French prison in central Hanoi, and now he was back—as our ambassador! At least one American was capable of forgiveness. In Hanoi's B-52 Museum (that's what it's called), featuring the exploits of the North Vietnamese Air Force, I saw Pete's fighter helmet on display, next to a piece of the wreckage from the fuselage of his F-4, his stenciled name still visible beneath what remained of the cockpit frame. Pete's story is an inspiration to all of us who lived through those war years; he retains no bitterness toward Vietnam or his own government. He even married a Vietnamese woman. Thanks to Pete and Senators John McCain and John Kerry, the United States finally normalized relations with Vietnam in 2001, President Clinton paid a visit to the country, and Hanoi ratified a trade pact with the United States shortly thereafter.

This should have made everything easier for us, but it had little effect.

James Wolfensohn, president of the World Bank, came to Vietnam for a look-see, as Vietnam was now the flavor of the month among international donors. I had met him earlier in D.C., after his big PV powwow, and invited him to see one of our solar communities in Vietnam. His money was helping to finance them. In Ho Chi Minh City, Wolfensohn met Shawn, who gave him a SELCO hat, which he wore the rest of the day to protect his head from the equatorial sun.

However, the World Bank, aside from our loan, never supported solar rural electrification in Vietnam, other than to finance study after study. In the end, the conventional power guys won out, and the first $50 million loan for rural electrification didn't contain one dollar for sun power. The bank provided heavily subsidized funds for unsustainable rural electrification that would never pay for itself, and the word got out that even the remotest regions would soon get electricity. Some did. We could compete with no electricity, but we couldn't compete with nearly free grid electricity. By 2011 the country had—by going deeply into debt to the World Bank (just like us!)—brought low-cost, heavily subsidized electricity to much of its rural population. But power generation would prove to be the next challenge.

In 2012 the government announced it was embarking on what may be an impossible dream: nuclear power and up to ten reactors by 2030! Well, it works for France, so who knows. But aside from risks and costs, there is still the distribution problem, and the best way to distribute power from the greatest nuclear reactor of them all—the sun—is through decentralized PV without wires.

SELCO-Vietnam was not a charity, and our alliance with the VWU was getting wobbly, since they much preferred the philanthropic, profitless, and entirely unsustainable SELF model of "spreading the light." VWU was a social service organization, and we were a business, and, alas, never the twain shall meet. Clearly, collecting money for a revolving solar fund they managed was one thing, but collecting deposits and monthly installments for a private company was a different kettle of *nuoc mam*. SELCO opted to go it alone, and I think the VWU was grateful. Nevertheless, in 2001 Madame Sam was flown—deservedly—to Vienna to a black-tie gala attended by half of Austrian high society, where she received the 2002 Energy Globe Award for bringing solar electricity to women in Vietnam.

For our eight-year effort fighting the Vietnam solar war, SELCO received a prestigious award of its own in 2001 from the U.S. Department of State. The award to the Solar Electric Light Company "for outstanding corporate citizenship, innovation, and exemplary international business practices in Vietnam" was presented at a ceremony at State's Dean Acheson Auditorium and featured a live video hookup from the U.S. consulate in Ho Chi Minh City, where SELCO-Vietnam staff simultaneously received a copy of the award from the U.S. ambassador. This was pretty exciting for our little company, to say the least.

Meanwhile, Harish Hande and Susantha Pinto, from India and Sri Lanka, were now making regular visits to Vietnam to help consolidate our far-flung operations and assist with overall management. Danh visited India to compare notes on operational methods and practices. Shawn and I continued to dream our impossible solar dream for Vietnam. So far, we were the only real player in the commercial solar energy field in Vietnam—now with some four thousand remote houses electrified—as well as the Siemens Solar dealer for the country. I went to Manila and got support for a massive solar initiative in Vietnam from the highest levels of the Asian Development Bank.

We lined up a half dozen provincial governments, which promised to pay half the costs of a solar subsidy themselves, recognizing that it would be cheaper to do this than attempt to distribute their costly grid power to so many remote hamlets. We had meetings with every chairman, minister, director, and chief of every agency, ministry, department, branch, and division of every level of government in Vietnam, from district to province to region to national.

But it slowly became clear that in a Communist state where no one is elected and there is no accountability, there is no need to make a decision if you don't have to. Officials loved meetings, but they hated making decisions, and when they did make them, they were always contingent on additional decisions being made higher up by faceless authorities who didn't like meetings and who loved to postpone their decisions. The process reminded me of the *New Yorker* cartoon of a businessman looking at his appointment calendar and speaking into the phone, saying, "No, Thursday's out. How about never? Is never good for you?" *Never* was good for the people who ran Vietnam.

Then, the World Bank arrived with $400 million to extend the grid to a thousand rural hamlets at a 90 percent subsidy, something

Satisfied SELCO-Vietnam customer in Bien Phuoc Province, Vietnam, with Tran Danh, company director. *(Author)*

bank officials in Washington had told me *they would never do,* since no country could pay back such a loan. This caused our provincial chiefs and their People's Committees to reconsider. Why should they pay for solar when Hanoi was going to bring them the grid for free, using World Bank funds? Our grand plans also suffered from dithering, procrastinating, and determined lethargy until SELCO-Vietnam finally back-burnered the big plans and focused again on day-to-day business.

We were in the solar sales and service business and should not have been a finance company, too; then we became a collection agency in order to meet the company's obligations to the World Bank, which graciously refinanced and rolled our loan over more than once. Vietnam constituted a huge and expensive learning curve for SELCO as well as for the World Bank. In Vietnam, no one defeats the Communists, not the full weight of the U.S. military over a decade of war; not globalization; not a booming economy, which is anything but Communist; and not SELCO. No one "beats" the Vietnamese at war or in business, as SELCO learned the hard way. They are the most stubborn people on earth, and they will have it their way, whatever that "way" may finally be. Maybe their way is the best way, and who can challenge that? It's their country.

On one of my last trips to Vietnam, newsstands everywhere were selling the *Time* magazine issue with former senator Bob Kerrey on the cover, illustrating the "Ghosts of Vietnam" feature story in which Kerrey's squad was reported to have killed twenty-one civilians in the hamlet of Thanh Phong in the eastern delta thirty-two years earlier. There were questions in the press as to whether Kerrey's actions were crimes, and he denied that he did anything illegal. All this was big news at the time and was covered in *The New York Times Magazine* and featured on CBS's *60 Minutes*. The Vietnamese were discussing it, too.

I talked about it with Danh when he got back from installing SHSs in the field. I mentioned the name of the hamlet, Thanh Phong in Tra Vinh Province, and asked if he had heard about the story.

"I was in that place yesterday," he said. "Thanh Phong. We just installed seven forty-Wp solar home systems there, in Thanh Phong. They all paid cash. I didn't know this story about Senator Kerrey."

Shawn returned to Los Angeles, where he sells real estate and dreams no more impossible dreams. Danh stayed on with SELCO-Vietnam; it became a smaller but independent company that today is the leading solar provider in the country, bringing solar power and light to those in need, those in the dark, wherever they are, providing they can afford it—without any government or donor subsidy. The company sells to cash markets, mostly, such as the army, national forests and parks, border patrol, and businesses, while manufacturing LED streetlights for domestic use and export (to Venezuela). A state-owned company, Red Sun, was set up in 2011 as a joint venture with a French company to assemble PV modules with cells from China.

SELCO-Vietnam can no longer afford to extend credit to poor people, and the government and provincial authorities, focusing on donor-funded grid extension, have written off the several million families that remain without electricity and are not likely to ever be connected to the national grid. Such are the politics of power in the two-thirds world, which you will learn more about in the following chapter.

Meanwhile, Vietnam remains one of the happiest, most peaceful and lovely nations on earth. And the war is over.

9

As the new millennium dawned, it seemed everyone was getting into the game of delivering solar energy to developing countries. Solar "projects" were proliferating, each one accompanied by three graduate students writing Ph.D. dissertations, at least five World Bank or Global Environment Facility analysts, and probably a dozen highly paid international energy consultants, plus filmmakers chronicling the efforts of those seeking to bring light to the poor. It was truly becoming a *cirque du soleil*.

The solar circus appeared in full force at a gala reception at the Swiss embassy in Washington, D.C., in the summer of 2000. The Swiss government had contributed a large chunk of money to a new $30 million fund for solar energy, the Solar Development Group (SDG), which was an outgrowth of our Pocantico conference. The "recovering venture capitalist" I had invited to Pocantico was the moving force behind SDG, which he incubated through his international development organization in Rosslyn, Virginia. SDG was well-funded by the Rockefeller Brothers Fund, the World Bank, and USAID and was staffed with highly paid managers, many of whom were at the Swiss reception: one hundred or so well-groomed men and women nicely turned out in dark business suits, the official Washington attire. Richard Hansen, formerly of Enersol and now of SOLUZ (see chapter 5), was also there with a new, respectable haircut and a trim sport coat. I was encouraged, ten years after I had founded SELF and four years since launching SELCO, that a commercial business development and investment fund was being launched to finance—exclusively—*solar power for the developing world*. Ten years earlier, I couldn't find anyone in Washington who could spell photovoltaics or who would believe that poor farmers in the two-thirds world could actually be provided with electricity from the sun. Now solar rural electrification was a veritable industry.

I was glad that international, low-cost, dedicated, public funds would be available, at last, to finance SELCO's growth and underwrite solar loan programs on behalf of rural people in the two-thirds world.

I listened to the Swiss ambassador give his introductory speech. He noted that two billion people, or 70 percent of the developing world, remained without access to electricity and that they continued to rely on a nineteenth-century light source, kerosene, which was dangerous, polluting, and unhealthy. He said that solar electricity had proved able to satisfy their demand for household electric power and that for millions of people around the world, solar photovoltaics offered the only solution to their world of darkness. Or some such. The speech was lifted almost directly from a SELF brochure I'd written a decade before. So it goes.

Meanwhile, the SDG eventually turned down nearly every project and every finance proposal we brought to it, arguing the risk was too great or the "return on investment" too slim or the business model wasn't "sustainable." Our equity investors had taken huge risks and were generally pleased with our progress, and our workers and managers risked life, limb, career, and a premature head of gray hair trying to make this difficult business successful. Yet SDG's unimaginative bureaucrats charged with using public and charitable funds from development organizations and foundations to finance solar business in the two-thirds world were afraid to take risks! Their jobs were on the line, of course, and the best way for a bureaucrat to avoid problems is to never, never, ever take a risk. In fact, if they avoided making a decision at all, if possible, or deferred all decisions for as long as possible, they would stay employed and comfortable, with an unlimited travel expense account at their disposal.

When the craziness of all this became apparent, I grew more cantankerous than ever as we found ourselves unsuspectingly walking the high wire of international development finance at the *cirque du soleil.* We never thought "international development" could be so complicated.

After all the fanfare, the SDG eventually folded and was liquidated. The group made few loans to solar companies in its first three years of operation. They were too busy "analyzing" potential investments, "investigating" possible recipients, and "processing applications" from solar entrepreneurs around the world, including, of course, our companies in India, Vietnam, and Sri Lanka. Most of the $30 million went back to the donors. Solar energy had apparently proved too risky. Investors were willing to risk their own money, but bureaucrats were afraid to risk *other* people's money.

However, the *cirque du soleil* was still in business. The Photovoltaic Market Transformation Initiative (PVMTI, mentioned in chapter 6 and announced at Pocantico in 1995) had by 2002 not yet made a long-promised loan to SELCO (we were "short-listed" in 1998). This meant nearly five years of salaries were paid and first-class airfares and accommodations covered as the capital managers and project consultants made their way from Washington and London to New Delhi and Bangalore on a regular basis. My observation that the United States entered, fought, and won World War II in less time than PVMTI took to make its first loan was not appreciated.

To understand the *cirque du soleil,* one must understand the Global Environment Facility (GEF), the five-billion-dollar fund set up by the World Bank and the United Nations to disburse money for environmental projects—including solar and renewable energy. Thousands of people were employed by consulting firms and project contractors through the UN, the UNDP, the World Bank, and the GEF to "prepare" and "implement" environmental projects. Some projects were successful and produced remarkable benefits, and most of these people were highly motivated and dedicated to their cause.

These global agencies were largely staffed by citizens of the two-thirds world who had made their homes in Washington and New York and who now drew down six-figure tax-free salaries. Many had become American citizens. Our apartment building in Chevy Chase was a veritable Tower of Babel, where hundreds of World Bank employees from India, Sri Lanka, Pakistan, Kenya, Egypt, Brazil, Turkey, Nigeria, and Indonesia lived out their comfortable lives, enjoying their careers as "development economists," with little intention of ever returning to what many saw as the godforsaken countries of their birth. Thus, it was likely to be a smiling, risk-averse, highly paid Indian bureaucrat in Washington who reviewed projects SELCO sought to develop with the World Bank or other UN agencies and then turned them down.

Meanwhile, in the nonprofit world, the *cirque du soleil* was an expanding franchise. Donor governments had gotten into solar electrification in a big way, financing huge solar-electrification projects in China, the Philippines, Bolivia, South Africa, and Indonesia. In many cases it was the strong arm of oil company subsidiaries like BP Solar, Shell Solar, and Total Energie that persuaded national aid agencies in Britain, the Netherlands, and France to support rural PV—because it benefited them! The largest was in China, where in 2003 the Dutch government made Shell Solar very happy by buying some 78,000 SHSs to be delivered, by Shell, nearly for free to peasant farmer and herder households in Xinjiang. Although these heavily subsidized projects were largely unsustainable, they did provide solar power and light to hundreds of thousands of deserving families who probably could never have afforded a solar lighting system.

China was to become the centerpiece of much of this activity as the United Nations Development Programme, and later the World Bank, picked up where SELF and our little Gansu PV Company left off. Although China had embarked on an ambitious program to bring the electric grid to thousands of remote villages, there were thousands more that it could not reach. I had trudged through many of them in the 1990s, when SELF installed over one thousand SHSs in Gansu Province, paid for by local commune funds and customer fees. SELF's Gansu project, which began with MaGiaCha in Tongwei County, was originally funded by the Rockefeller Foundation. It was the catalyst for all that came after and the incubator of dozens

Chinese peasant farmer cleans his new 20-watt solar module in Tongwei County, Gansu Province, China, in 1993—twenty years ago! *(Author)*

of new solar companies, government programs, and international-donor-aid projects in China.

The heart of SELF's China project was the Gansu PV Company, the Sino-American joint venture of Professor Wang Anhua's Gansu Natural Energy Research Institute and SELF (see chapter 2). Headquartered at its fifth-floor offices in central Lanzhou, GPV had perfected a "cash-and-carry, plug-and-play" 20-Wp portable SHS that required no service network to install or maintain. The Panasonic deep-cycle batteries were housed in a serviceable red wooden box that contained the charge controller, switches, and easy-to-read meters. Professor Wang also designed, built, and installed systems for remote forest-ranger stations, police outposts, and telecom transmitters. As a result of our efforts setting up the first Sino-American solar joint venture, we had inadvertently launched a Chinese *cirque du soleil*. Everyone wanted to join up. Soon, I found myself speaking at numerous PV conferences in Beijing, while Chinese officials made their way out to Gansu to see MaGiaCha in Tongwei County and to visit Professor Wang at GPV.

GPV had its biggest success selling portable solar systems to Tibetan nomads, who grazed their yaks on the vast grasslands in southern Gansu Province. There are actually more Tibetans in China (in Qinghai and Gansu) than in Tibet, and they are left alone to pursue their nomadic lifestyle. Even their famous, remote monasteries were flourishing, and Tibetan monks at Labreng Monastery in Huining County eagerly sought to become distributors for GPV. They sent young monks to Lanzhou for PV training at GPV's headquarters. Soon, GPV was delivering solar lighting systems to rural western China by yak-drawn carts.

One happy herdsman was featured in a CNN report, standing outside his solar-powered yurt. SELF's Bob Freling, who replaced

A Tibetan family in Gansu Province, China, is all smiles with their sun-powered light and radio in their yurt. *(Robert Freling)*

me as executive director in 1997 when I left to launch the Solar Electric Light Company, fought his way through the foreign-affairs bureaucracy in Beijing and Lanzhou to get permission for CNN's crew to travel to the Tibetan areas of Gansu. The report was seen around the world on CNN.

Professor Wang focused on marketing his solar systems in Gansu through weekly radio "infomercials" broadcast province-wide. Lack of credit remained a problem; as GPV discovered, Chinese peasants don't do credit. Neither do banks, which in China's wild west were mostly insolvent and corrupt. This is why the Tibetan market was so important: nomads had money from selling livestock, and they paid cash. They also knew no government program would ever connect them to the grid since they moved around, as nomadic people tend to do. They were perfect candidates for portable solar lighting systems! Our farmer market, on the other hand, was very difficult to serve. Peasants remain among the poorest people in China because the government sets agricultural product prices to benefit cities, not the farm economy, and because corrupt local officials heavily oppress farming communes every way they can. Without credit, or heavy subsidies, farmers could not afford solar. SELF could subsidize SHSs, and so could the government, but GPV could not.

In any case, GPV was now famous. A Beijing taxi driver told me he had even seen a story about it on national TV. Back in Washington, I would bump into colleagues who had just returned from China to study solar, and they too had ventured out west to industrial Lanzhou to visit China's famous pioneer of solar electricity, Wang Anhua. American solar experts issued reports that featured GPV and monitored a growing industry comprising some four hundred small solar companies—imitators of GPV—in the world's most populous nation. The Darden School of Business at the University of Virginia did a case study on SELF in China in 1997. It reported that SELF and GPV had been important drivers of solar rural electrification in China.

By 1999 the Gansu PV Company had installed some seven thousand small SHSs—a drop in the bucket, really, but more than any other company had done thus far. Most of GPV's sales came from peasants who heard the company's radio ads; took a bus from their remote communes in rural counties; found their way to the office in the huge, dirty, provincial capital on the banks of the Yellow River; and walked up the five flights to make a cash purchase of an SHS. The product was explained to them—how it worked and how to install it. Then, it was boxed up, shrink-wrapped, and hoisted onto their backs, and off they went as solar pioneers in their own right, the first in their town with electric light.

Customer lists became the company's marketing program, and every customer was recruited to bring in more buyers; prizes and incentives were offered, mirroring the Amway sales approach. Peasants became GPV brokers. Testimonials were the preferred form of advertising, used on the radio and even in several television specials sponsored by GPV. Marketing "collateral," GPV included tens of thousands of pamphlets, flyers, and slick color brochures describing its products. Professor Wang handed out promotional cigarette lighters with the company's logo because without the always-lit kerosene bottle lamp on the family table, the men had no way to light their cigarettes. This was the only downside to owning an "Anhua SHS."

Letters poured in from people who wanted to buy Professor Wang's SHS, and Bob Freling, fluent in Chinese, translated them:

Mr. Wang,

I am a farmer living in a remote village on Wushan Mountain. Every time I see people in the valley below with lights in the villages, every time I see them watching TV while they have their supper, I become very envious. Although we have enough to eat in this village, our sources of income are still rather limited. At present we have no way of coming up with enough cash to bring a power line to our village. Therefore, I am thinking to use my own money to purchase one of your SHS.

Respectfully, Wang Shucun

From Liu Chunming, Dingxi County:

Hello! In the small town where I live, we often suffer from electric blackouts. For example, we completely missed the Hong Kong Handover on July 1 [1997] due to a blackout. Several days ago, I heard a radio advertisement discussing a solar device that can "borrow the sun to light up the night."

Zhou Xiangyi wrote from Ming County:

I am a farmer living in a mountain village. I was recently appointed head of my village. The other day I heard a radio announcement discussing how your "Anhua" solar home system can provide light to unelectrified villages. This news made everyone in our village very excited. We would like to purchase thirty SHS.

The writer then proposes that they make a down payment and then installment payments "until the units have been completely paid for. This is a poor village. We do not have much money."

And from Nan Lianxi, in Qingshui County:

After hearing about your solar home systems from the radio advertisement, I became very interested. All fifty homes in our village are without electricity. At night we have to live under the dim light of kerosene lamps. Many farmers in our village have not even had a chance to hear about solar electricity, so it is difficult for them to become consumers.

This was a huge untapped market, with little money and no credit. Nonetheless, thousands of families did buy solar, and the testimonials are a joy to read:

"I wanted to watch TV before I died," Luo Yanzhen told a reporter from the *Far Eastern Economic Review*. A GPV customer from a village 180 kilometers southeast of Lanzhou, she also bought a black-and-white TV to go with her small PV system. "I've seen Premier Li

Peng and President Zhang Zumin for the first time." The weathered farmer had sold ten small sheep and eight baby pigs to purchase a solar home system, "but it was worth it," she said.

Over the years, I read or listened to hundreds of personal testimonials from peasants as they described how electricity changed their lives, how they no longer needed to use kerosene, how much better their children could study at night, and how much they enjoyed seeing television for the first time. One I recall was an old man, who said, having seen the Chinese premier on TV, "I have never before been able to see the emperor!" Another elderly peasant, shedding tears, said, "I have long heard that city folks do not need oil to generate light, but in all my seventy years, this is the first time to actually see such a phenomenon with my own eyes. What a beautiful thing!"

One customer wrote to Professor Wang after installing an SHS:

As the afternoon wore on, the house filled with more and more neighbors from the village, who continuously turned on and off the SHS and asked all sorts of questions. Expressions of envy as well as expectation were clearly visible on their faces. At that moment, their hearts were illuminated with the light of hope. The next day, another family in Xiping village decided to purchase a solar home system. They have a son who is about to take an examination to enter high school. The parents hope that the bright light of the SHS will help their son better prepare for the exam. The parents have invested a great deal of hope in their son, and now they are also putting hope in solar electricity. Suddenly, I feel as though this year's sun is especially bright, and this year's winter especially warm.

But the best came from Liu Xueming in Huining County:

On the afternoon of October 4, I returned to my village, carrying a television set, radio, and a large cardboard box . . . everyone was staring at the mysterious box I was carrying. I removed the solar panel from the box and connected the positive and negative wires to it as indicated in the instructions. Immediately, a green LED light lit up as the wires were connected. Gasps of amazement came from the room. How could a glass panel turn on a green light, just like that? I laughed and explained the green LED indicated that the battery was being charged by the solar panel. I told everyone to be patient and that in a few minutes they could watch television. As I turned to connect the TV set, I could hear a few older people murmuring among themselves. "Don't be so boastful! How could a small glass panel, exposed to a bit of sunlight, allow us to watch television? Don't pull the wool over our eyes just because we're a bunch of old country bumpkins!"

I proceeded to turn on the TV set. Everyone became very excited. They hugged one another, and shouted with glee, "We have electricity in the village! We can watch television!" My own parents asked everyone to take a seat and get comfortable. The older ones sat on the sofa, the children sat on the kang. Everyone sat there watching TV with huge smiles on their faces. My wife poured tea for the older ones. We watched a soap opera, and when it was dark I turned on the electric lights, and again exclamations of surprise filled the room. "How much brighter than our kerosene lamps!"

My son was born a few weeks after I brought home our solar home system. Our home shone so brightly that night! As relatives and neighbors came by to congratulate us, I could not help but reflect on my good fortune. Everyone said to me, "Xueming, you really pulled it off. You went to Lanzhou and came back with lights and a television! This is something we have never seen in all our years." In honor of our good fortune, I decided to name my child Guang Dian.

"Guang Dian" means "photovoltaics" in Chinese.

I considered bringing GPV into the SELCO fold, since we had acquired SELF's 49 percent ownership of the company. But despite the booming economy in Gansu, the stability of the yuan, the right of foreigners to now own more than 51 percent of critical industries like energy, plus the constitutional recognition of private enterprise, the newly official status of capitalists in China, and all the glowing reports about GPV, I remained skeptical about the wisdom of investing SELCO's capital in GPV. I wasn't getting clear information from the company.

Bob and two other Mandarin-speaking Americans tried to sort things out, but it seemed to me that many Chinese were actually offended when foreigners dared to learn their language, making it impossible for them to talk behind our backs, make fun of the "long noses," and otherwise scheme against our interests. Professor Wang often became incensed when I brought my own Chinese-speaking colleagues and independent translators to meet with officials, instead of using his, since he could not "correct" the person doing the translating. Fortunately, Professor Wang spoke just enough English so that he and I could communicate directly, and his communication skills improved considerably during rounds of highly potent maotais at the nightly banquets. (He learned his halting English while under house arrest during the Cultural Revolution of the 1960s.)

My concern was that GPV was not growing at the rate the market warranted, despite the credit issues. Quarterly financial reports, with official stamps from the provincial tax authorities, always showed losses. Professor Wang made lots of sales, but revenues never exceeded costs, although I knew his thirty-six employees were not costing him that much, and he owned his workshop and offices. Copycat companies had sprung up on Professor Wang's doorstep—which he tried hard to hide from me. Many were selling technically inferior products. Several former employees, as well as people he had trained earlier at the UNDP-funded Gansu Natural Energy Research Institute, started their own small firms. Because they were younger, less academic, and more businesslike, they soon had large PV concerns of their own, selling to the huge cash market of nomadic herders in Qinghai Province next door to Gansu.

As Professor Wang explained to me one day, "The largest social institution that works in China is the family." He trusted no one outside his immediate family. This same phenomenon exists in Vietnam, where there is an economic wasteland between family-owned businesses and state-owned enterprises, or between small, local companies and foreign offices of multinational corporations; nothing in these societies is transparent.

A half-million-dollar program funded by the National Renewable Energy Laboratory provided subsidies for farmers purchasing PV systems, until the money ran out: this was SELF's biggest government contract to date, a cost-shared project between the U.S. Department of Energy and the Ministry of Science and Technology in Beijing. It was facilitated by William Wallace, whom I invited to visit Gansu, and who later headed up the UNDP's renewable energy office in Beijing, where, with the authority of the UN behind him, he began talking to the Chinese government about integrating solar photovoltaics into its rural and agricultural programs.

All along, I was meeting with World Bank staff in Beijing and with colleagues at the Asia Alternative Energy Unit back in Washington. Streams of World Bank, UNDP, and GEF consultants made their pilgrimage to GPV to see how this little company was able to manufacture and sell such small solarelectric systems that, thanks to Wang Anhua's electrical engineering skills, were not only reliable but also amazingly efficient. Despite everyone's admiration for GPV's pioneering efforts, the government and the donor agencies preferred to support the aggressive new upstart PV companies run by young Chinese entrepreneurs. Professor Wang, stubborn and independent and distrusting of "officials," was his own worst enemy; he was unable to keep up with the dizzying dynamics of a changing China. GPV continued to burn through its capital.

Meanwhile, the World Bank's legions of energy economists had been busy, and in 2002 the bank put $205 million into the China Renewable Energy Development Project, a quarter of it earmarked for PV, to partially fund some 350,000 SHSs with a $1.50-per-watt subsidy. I took comfort, at least, in knowing that SELF and Professor Wang had started all this at MaGiaCha, China's first solar village.

Eclipsing the World Bank project was the Chinese government's own ambitious Chinese Brightness Program, which aimed to deliver solar power to one million peasants in Tibet, Yunan, and Sichuan. This was jointly funded by China's Ministry of Finance and KfW, the German state-owned development bank. Too late for GPV.

Our SELCO board agreed that China was a bad bet for us. John Kuhns, whom you met in chapter 5, had earlier been a pioneer investor in China, buying operating hydroelectric plants around the country, a picaresque tale well told in his colorful 2011 novel, *China Fortunes*. The book, a thinly veiled autobiography, clearly depicts the chicanery and intrigue inflicted on anyone doing business in China, then and now.

John originally thought we could make money selling solar in China, but later thought better about it. What neither he nor I could have imagined at the time was that one day China would be selling solar to the world! I had visited the primitive PV factories in China, built with government money, and while they manufactured small solar panels perfectly suited to our rural markets, they were not of export grade. As you'll see shortly, that was to change. I could not have foreseen that, only a few years later, the next company I started would be buying millions of dollars' worth of solar panels from China!

John Kuhns returned to China in 2006 to launch—with a $50 million investment fund—the China Hydroelectric Corporation, which bought or constructed nearly five hundred megawatts of small hydro installations around this mountainous land and made money selling electricity to the national electric grid. These hydro resources, coupled with the completion of the Three Gorges Dam on the Yangtze and compounded by a huge push to build coal-fired generating plants (one a week!), alleviated the booming country's power shortages, and soon the grid reached even some of the remote hamlets we had earlier electrified with our solar systems. The aforementioned solar programs focused on the remaining 5 percent of the rural and herder population that was never going to see the grid, still a sizable market of over 50 million people.

After SELCO in 2000 decided against investing in GPV, Wang Anhua fell on ill health. We looked at partnering with some of the new solar players in China, but they all told me they didn't need any Western capital. And no one wanted to buy GPV or its technology.

Not when they could steal it (no copyright, patent, or other proprietary protections are available in China, where much of "capitalism" might better be described as rampant piracy). The company, exhausting its capital, eventually faded from view, and Professor Wang retired to Sweden to live with his daughter. His was a life well spent, and he will always be recognized as the true pioneer of solar electricity for the rural people of China.

The Chinese government announced at the 2004 World Renewable Energy Conference in Berlin that it would meet 10 percent of its energy needs from renewable resources by 2020. "Priority will be given to developing renewable energy in rural and remote areas to meet the lifestyle and work demands of local people," the government stated. SELF had arrived in China ten years too early.

The State Planning Commission now had a "renewable portfolio standard," requiring a substantial percentage of provincial power generation to come from renewables, including solar power, something some U.S. states, like Florida, have still not managed to do. The government made funds available to China's many solar-manufacturing facilities, mostly owned by one or another government ministry. And in 2004 the State Development and Reform Commission announced its own program of "solar energy exploitation" to the tune of $1.4 billion. *China Daily* had earlier reported that severe environmental problems resulting from the country's reliance on coal, coupled with serious energy shortages, were driving the National People's Congress to consider massive support for renewable energy. This new direction on energy policy would have enormous consequences.

By late 2004, China was experiencing near-catastrophic energy shortages and expecting a power shortfall of 30 million kilowatts. At the same time, the country's skies were perpetually hazy from burning carbon fuels. Some of these particulates have been identified falling on California. (In addition, between six thousand and ten thousand coal miners died each year in the mines.) By the end of the decade, China had become the world's second-largest economy, and the world's largest producer of CO_2.

Napoleon had observed, "Let the Chinese giant sleep, for when it awakens, the world shall tremble." We isolated Mao and his Communist revolution, but we can't isolate Chinese capitalism. China will soon become the largest user of energy in the world. What they do there affects us all. One ray of hope is that by mid-decade the government decided to embark on the most ambitious domestic renewable energy program in the world. Out west, Gansu alone was spending 40 billion yuan on renewable energy; thirty-six electric windmills were being installed every day by 2012; a 200-megawatt "solar park" at Golmud was among the largest in the world; while other large utility-scale PV plants were under construction near Dunhuang. Province official Song Rongwu said, "Ten years from now I believe every home in Dunhuang will be powered by clean energy. The Gobi desert will be filled with blue photovoltaic panels. It will be a beautiful sight."

I first visited China as a guest of the Chinese government in 1979. Having studied Chinese history in college, and having read Jan Myrdal's *Report from a Chinese Village* and Edgar Snow's *Red Star over China,* I was fascinated by the age-old glories of the Middle Kingdom and its successor, the People's Republic. For the *Colorado Daily* in 1966, I wrote an award-winning column advocating—horrors!—the recognition of "Red China" by the U.S. government, an event that was still ten years in the future. The China I saw in 1979—with giant pictures of Marx, Engels, Lenin, and Stalin dominating Tienanmen Square—can only be evoked, in summary, by imagining a time traveler going back to America in 1920. Or imagine a child of ten watching the *Wright Flyer* take off from Kitty Hawk in 1903, and then at age sixty-six flying to Europe in 1959 on a Boeing 707. Now, compress a comparable half century of changes such as we have known in the West into ten and you've got the pace of development of modern China. This phenomenal economic growth is quite beyond comprehension, but there it is. If someone in 2000 had told me that, ten years later, Chinese would be buying more Buicks than Americans, I would have responded with, "Yes, and in ten years I expect to be able to travel on my personal rocket ship to my vacation home on the moon."

China will soon surpass the West in many endeavors, and has already done so in the production of solar cells and modules. It is on a rapid march toward being the world's biggest user of solar energy, from solar electricity to solar thermal to concentrating solar to solar cookers and household solar water heaters, becoming what has been called "the first green superpower."

The modern PV-manufacturing industry in China can be told in the story of one man: Zhengrong Shi. Born in 1963, the son of destitute farmers in Jiangsu Province, Shi managed to get into a Chinese university and earn a degree in physics. In 1988 he became a foreign-exchange scholar at the University of New South Wales in Australia, where he met the renowned solar scientist Martin Green, who entered him into a doctoral program in solar-cell research. Twelve years later, Shi noticed the enormous growth of the economy in his former homeland. He decided China would need new forms of clean energy, and he left full-time research to return home and start a solar-photovoltaics factory. He chose Wuxi, which I visited in 1979 when it was a sooty provincial town with dirty canals and ancient stone buildings, none over two stories; today it is a progressive, modern economic powerhouse that gave Dr. Shi $6 million to start his company, which he named Suntech.

As a solar technologist, Dr. Shi knew how to create an efficient production facility—from solar cells to high-quality finished modules—that could compete in the world market, unlike the primitive Chinese PV panels we'd purchased at GPV only a few years earlier. Launched in 2001, Suntech began shipping solar panels to Germany to supply its renewable energy program (see following chapter), and business exploded.

Dr. Shi found international investment bankers to buy out the Wuxi government's share, and in 2005 he took the company public on the New York Stock Exchange, raising $400 million in its first initial public offering (IPO). Paralleling the massive growth of China itself, Suntech quickly grew to become the world's largest manufacturer of solar panels, surpassing Japanese companies like Sharp, Kyocera, and Mitsubishi. Thanks to Dr. Shi, Suntech was able to increase panel-conversion efficiency to 22 percent, the highest of any commercial solar module at the time. "Solar will be cheaper than coal or gas," he told *Fortune* magazine. In 2007, *Time* magazine named Zhengrong Shi one of its "Heroes of the Environment." By 2010, Suntech had five thousand employees and a production capacity of 1,800 megawatts, with revenues of well over $2 billion; it had shipped over 13 million solar panels to more than eighty countries.

Few American Dream stories can top Dr. Zhengrong Shi's journey from Chinese peasant boy to the country's first renewable energy billionaire, who is reported to be worth $4 billion today. I met Dr. Shi at the National Press Club in Washington in 2008, and he said he wanted to see China use solar more, and not just export its PV production to the West. He also said he supported efforts to enable poor people without electricity to use solar, but clearly supplying the off-grid market with small 20-Wp and 40-Wp panels was not in the interest of Suntech, which was manufacturing 240-Wp panels for export.

The overnight success of Suntech prompted the emergence of new Chinese solar ventures, which began springing up faster than we in the West could count: Yingli, Trina Solar, Jinko Solar, Ja Solar, Solarfun, and Canadian Solar (owned by overseas Chinese from Ontario), to name a few. At a solar PV conference I attended in 2011, more than thirty Chinese solar companies were exhibiting, most of which we'd never heard of. These companies were supplying the national solar programs in Germany, Italy, Spain, and Portugal but now recognized that North America was the next big market for solar PV. With production costs a fraction of those in the West, Chinese PV companies were causing havoc on world markets by 2011, and a trade war ensued (see chapter 12). Even China revealed itself to be what Greentech Media called a "destructively dynamic" solar market when Dr. Shi was removed as CEO of Suntech for taking on too much debt leading to bankruptcy in 2013. The City of Wuxi assumed the debt and took over the company.

"When I started Suntech in 2001," he recalled after stepping down, "the average levelized cost of energy for a solar project in good sunlight was around fifty cents a kilowatt-hour. Today, new solar power plants are promising clean electricity for less than ten cents per kilowatt-hour, without subsidies."

By the end of the decade, China was the world's largest emitter of greenhouse gases, surpassing the United States. The Chinese government had become concerned about their environment, worried about the terrible pollution caused by the country's ever-expanding use of coal. Officials were irked that 90 percent of the country's solar PV shipments were going overseas. So, in 2011 China's National Energy Administration (NEA) announced it would set a goal of installed-solar-power generating capacity of 15 *gigawatts* by 2015. Given that they built a high-speed rail system faster than we built our interstate road system, they will probably do it. One reason to push the domestic market is that all these new solar companies created an overcapacity of production to supply a world market where PV panel prices subsequently had dropped 51 percent by 2012 because of oversupply. If the Chinese will buy Buicks made for domestic consumption in China, they will buy solar panels made in China.

The government recently announced plans for a national solar program such as the one instituted by Germany in 2000 (see chapter 12). China's NEA is also looking to foreign investors to build utility-scale solar projects and help develop the country's transmission infrastructure to serve renewable energy (solar, wind, hydro). China needs energy, and the potential market is huge if it is to catch up with the West. Its annual per-capital kilowatt-hour consumption, according to John Kuhns's China Hydroelectric Corporation, is one fifth that of the United States. If it catches up using coal, we are all in trouble. The hope is that distributed sun power and other renewables supplying an expanded, efficient grid in China will play a significant role. Besides utility-scale central PV plants, the biggest future market of all are decentralized grid-tied, net-metered households and small commercial solar systems, which have driven the Western solar business but have not yet been introduced to China. The good news is that China had already installed 7.4 gigawatts of solar by 2012 (compared to India's 145 megawatts).

Dr. Shi, who became known as "the Sun King," said at the World Economic Forum in Davos in 2012, "Solar is getting so cheap that we believe that half the countries in the world will reach grid-parity by 2015." This means solar PV, without subsidies, will be competing head-on with conventional fossil-fuel-generated power. Dr. Shi said PV power markets are moving away from Europe and are set to expand rapidly in China, the United States, India, Japan, and South Africa.

The dream of harnessing power from the sun was becoming a reality, and the *cirque du soleil* would soon become the greatest show on earth!

10

I first met Dr. Hermann Scheer in Harare, Zimbabwe, in 1993, where he and his "co-conspirator," Dr. Wolfgang Palz, were hosting a German-sponsored conference on solar power for the developing world. I was there with the UNDP mission to design a solar-electrification project for the country. Scheer—to whom this book is dedicated—was a member of the German Bundestag, and Palz, a pioneering PV physicist who had worked in semiconductor research since the late 1960s, headed the European Union's Solar Programme. These are two of the most remarkable men I have ever had the privilege to know.

The late Dr. Scheer was busy promoting solar policies in Germany, while Palz had been pushing solar in Europe for twenty years. They had been promoting solar for the world's poor under the banner "Power for the World," believing that the two billion people on earth without electricity were entitled to a minimum of 10 watts apiece. This would mean a 50-Wp solar panel for a family of five, not an unreasonable idea if the EU, USAID, the United Nations, and the World Bank would get behind it. The trillion dollars America spent on the Iraq and Afghanistan wars alone could have done the job, but I'm not naive enough to think swords will ever be beaten into solar panels. Yet, like Palz and Scheer, we pursue the dream nonetheless, and along the way we all surprise ourselves with what we, ever the "believers," do manage to accomplish. Palz and Scheer helped lead the way, and since our meeting in Harare and for the twenty years following, they never ceased to inspire me as we crossed paths around the world at endless solar conferences. (You'll recall Wolfgang, as I came to know him, accompanied us on a return visit to

Pulimarang, SELF's solar village in Nepal, where he was deeply moved by how dark a rural village with no electricity can be and how much light a small solar panel could provide.)

In 2011 Wolfgang Palz published a landmark book called *Power for the World: The Emergence of Electricity from the Sun*. With an introduction by Scheer, and 140 pages of his own solar diaries, the 575-page book includes writings by forty-one "solar pioneers" Palz corralled from around the world. It covers the history of solar electricity, the research and the technologies, the development of the industry, worldwide government solar programs, and the global solar business today, all written by career experts in PV. No one truly interested in solar energy can be without it any more than a clergyman could be without a Bible.

I was honored to be asked to contribute a chapter, and I provided a short update on the activities of SELF and SELCO. With an office on Washington's K Street and under the leadership of Bob Freling, SELF now concentrated its efforts on South Africa, Brazil, Nigeria, Bhutan, Kenya, Burundi, Rwanda, Benin, Lesotho, and Haiti. Bob, the most unflappable and indefatigable traveler I've ever known, crisscrossed the globe initiating solar projects wherever there was a need and local partners to work with. He would then find the necessary funds.

Bob coined the phrase "Energy Is a Human Right," gave inspiring speeches from Paris to Portland and Amman to Aspen, and persuaded wealthy donors, foundations, and the new breed of "entrepreneurial philanthropists" that was emerging from Silicon Valley to support SELF's projects. I was sitting in his office one day in 2010 when he opened his mail and out fell an "over the transom" check for $500,000 from an anonymous donor.

Bob attracted some wealthy and influential supporters to SELF's board, including actor Larry Hagman, Templeton Prize–winning physicist Freeman Dyson, Paul Mitchell hair products CEO Jean Paul DeJoria, and San Francisco philanthropists Steven and Mary Swig. Ed Begley Jr., the actor at whose Studio City house I had held a SELF fund-raiser in the early days, joined the board. Ed was everybody's environmentalist; he actually lived in a solar-powered house, cooked vegetarian meals on a solar cooker outdoors, and drove an electric car. With a few more Ed Begleys around, the world would be a better place. Larry Hagman (who left us in 2012) held fund-raisers at his mountaintop aerie in Ojai, California, that was powered by over 80 kilowatts of PV panels, and Freeman Dyson opined for the national media on how SELF was changing the world. Not exactly true, but a nice sentiment. None of us was really changing the world, only changing ourselves and helping a few folks along the way as best we could.

Following on the solar-electrification project I'd begun many years before in South Africa's Valley of a Thousand Hills, Bob helped the Zulu community of Mapaphethe close the "digital divide" with a pilot solar project centered on Myeka High School. Solar power supplied by SELF allowed the school to operate twenty-five computers contributed by Dell-SA, while SELF provided a satellite uplink and receiving station powered by the sun. Students could finally access the Internet in a community that, until then, had only one solar-powered telephone. Soon, they had their own Web site and were "plugged in" to the global solar-powered schools program, which looked to Myeka as a beacon of light and hope. Two students from Myeka High School later were selected by the International Solar Energy Society (ISES) to present their school's solar project at the annual ISES conference in Mexico City.

SELF had been instrumental in getting the Mandela government to focus on SHS as a complement to the aggressive rural-electrification program by the country's power utility, Eskom. The government announced plans for a 90 percent subsidy for SHSs and proposed that two million homes beyond the reach of the grid would be targeted for solar—and then nothing happened for five years, aside from a failed program instituted by Shell and Eskom. Private solar companies subsequently stepped into the breach.

SELF electrified a coastal community in Ceará, Brazil, north of Fortaleza, in the 1990s, and in 2002 SELF went back to Brazil to bring electricity to a village located in the Xixuaú-Xiparíná Reserve, forty miles up the Amazon from Manaus. Here, SELF installed a solar power plant big enough to operate a large satellite dish, computers, lights, and a vaccine refrigerator. The Caboclo people of this remote Brazilian rainforest preserve could access information through the Internet and communicate with the outside world for the first time. Bob brought along a video crew to chronicle the project, producing an excellent documentary for television.

Bob then took SELF into one of the toughest nations to deal with in the world—Nigeria. Africa's most populous country (130 million people) has three things besides lots of people: oil, corruption, and a lack of electricity outside the cities. And it has sunshine. The governor of Jigawa, a state in the dry Muslim north near the border of Niger, asked SELF to bring solar power to three villages that had never seen electric light or electricity. USAID and the U.S. DOE stepped in, matching funds from the state government, and SELF launched a half-million-dollar pilot solar project, the largest of its kind ever attempted in Nigeria. The governor proudly inaugurated the project in early 2004, with a CNN news crew capturing on video the delight of villagers, who now had a source of power to run a water-pumping system, mobile irrigation pumps, streetlights, lighting and computers in their schools, a vaccine refrigerator for their health clinic, and lights for surgery rooms and for three mosques. A 1.6-kW array powered a unique microenterprise center in a one-story cinder-block building, where six small businesses leased space. Sixty householders also signed up to purchase SHSs. All this was powered not by oil, but by the blazing, relentless African sun.

CNN ran the story for a month worldwide. President Obasanjo, whom Bob had met in Washington, D.C., visited the project. Three other Nigerian states asked SELF to help them do similar projects. One hopes that Nigeria will plow its oil profits into bringing solar power to its people, who have gained little from selling off the nation's wealth. Why should foreign donors have to do this when Nigeria is so rich?

In the Himalayan kingdom of Bhutan, SELF found a partner in the Royal Society for the Protection of Nature, which helped subsidize the cost of 151 four- and six-light SHSs in a nature preserve for endangered black-necked cranes because installing overhead electric

Young women connecting PV panels for SELF atop home in unelectrified region of Bhutan. (*Robert Freling*)

lines would be hazardous to their health. Not having electricity was also hazardous to the health of Phobjikha's residents, since they had relied on kerosene lighting for the past century, replaced by wood when they could not get kerosene. SELF installed a 750-watt PV system at the nature center's headquarters to run a computer and other office equipment.

In a cooperative venture between the two organizations I had started, which had long since gone their own ways, SELCO-India was contracted to supply all the systems, do the installations, and send technicians to train seven local youths in solar-system maintenance. The equipment and materials were shipped by truck from Bangalore all the way to Bhutan. Following SELF's modus operandi of not giving away systems, as so many failed donor programs had done, and instilling a sense of responsibility ("SELF-help"), the villagers in this remote hamlet were asked to pay part of the cost of their SHSs on a three-year installment plan. "I can pay for the installments from the sale of potatoes," said Tshewang Zam when she turned on her lights for the first time. "Besides that, we can also stitch our embroideries for sale, we can do weaving at night, and our children can study under better conditions."

In West Africa's Benin, Bob launched his Solar Market Garden Program, featuring an innovative irrigation system that enabled women farmers to grow high-value fruits and vegetables year-round. Bob's new vision for SELF was to fund "whole village" solar-based development, including village water systems and solar-powered drip irrigation. "The garden saved me. I'm always in the garden. We didn't know that the sun could do all of this. Now, we sell, we eat . . . we eat a lot here!" said Ganigui Guera, president of the local women's farming collective. A French television crew produced a feature story on the project.

SELF teamed up with Dr. Paul Farmer's Partners in Health organization to electrify AIDS treatment centers and health clinics in Rwanda, Burundi, Lesotho, and Haiti, with support from USAID, Google, and the Clinton Foundation. Bill Clinton flew into Rwanda to see the AIDS clinics, where the sun provided power for vaccine refrigerators, lights, computers, and diagnostic equipment. A photo of a smiling Clinton with his arm around Bob soon arrived by e-mail. In February 2012, President Clinton visited SELF's latest project in Boucan Carré, Haiti: twenty-one schools, a microenterprise center, a solar market garden, and two fish farms, totaling 200 kilowatts of PV installations funded by NRG Energy through its one-million-dollar commitment to the Clinton Global Initiative. Bob and his project manager, Jean-Baptiste, led President Clinton on a tour of the sun-powered health center, took him into the battery room, and explained how the inverters worked. President Clinton's interest in solar energy never flags.

Solar technology doesn't just show up in all these remote parts of the world where it can serve its highest and best use: someone must bring it there, as SELF and SELCO and many others do, and then someone must install it. This means trained technicians are required. You can organize a project to benefit willing recipients, and donors can write all the checks they wish, but people have to do the work, and they are my heroes.

One of my biggest heroes is Johnny Weiss, who founded the non-profit solar-energy-training organization Solar Energy International (SEI) in Carbondale, Colorado. Launched in 1991—just after I had founded SELF—to teach "sustainable energy" for international development, it offers beginner, intermediate, and advanced courses in solar photovoltaics (and other solar technologies). I signed up immediately, flew to nearby Aspen, and spent a week learning the difference between alternating current and direct current and all about the mystery of watts, amps, and volts. We learned how PV worked, or, rather, what it does, how much power a certain size panel produces, how much battery storage is required, what kind of "loads" it can support, et cetera. I took the test and passed the course, and over the years forgot most of what I learned. I would leave the technical specifics to people a lot smarter than me.

SEI is now in is twenty-second year, with a staff of twenty-eight at its two educational centers. It has sixteen hundred "alumni" from fifty states and sixty-six countries, many of whom are active in international solar work or who have become executives in commercial solar companies. SEI is not only the largest but the oldest PV training organization in the world. SELF hired Johnny in the early days to train electricians in the Solomon Islands and China. Today, SEI runs training courses all over the world. In 2012 SEI published the ultimate technical manual for on-grid solar, the *Solar Electric Handbook* (see Bibliography).

A fellow SEI alumnus and a wiry vortex of energy, Walt Ratterman, came to see me one day at the SELCO headquarters in Chevy Chase. An electrical engineer and a licensed electrician in three states, he then owned a 110-man electrical contracting company in York, Pennsylvania. He also had a fascination with solar PV. After taking the SEI training course, he earned a master's in renewable energy from an online university. He had learned about SELF and SELCO and came to offer his help. "What about your company?" I asked him. "I'm closing it down to work full-time on solar," he replied. He didn't want to be paid, only to be part of the global solar mission. Before we could put him to work, however, he volunteered for international relief work through Knightsbridge International, which delivered food aid to remote regions suffering from war and other calamities. Knightsbridge, profiled on an ABC television news special, introduced him to the need for solar power in the two-thirds world.

While on a mission to bring a convoy of donated food from Uzbekistan to northern Afghanistan, Walt found himself on the front lines of the Northern Alliance fighting the Taliban in 2001. He always traveled with a solar-powered Iridium satellite phone and video camera, and one day he sent me a live uplinked video feed of explosions, just a few miles from his truck convoy, showing bombs raining down from B-52s days after the United States launched its attacks. Not only was Walt a technical genius, he was a concert pianist and a wonderful father of two talented children, and he lived off the grid in a solar-powered house. We hired his son, Shane, at SELCO one summer.

Learning what he could from SELF and SELCO, Walt started his own nonprofit organization, SunEnergy Power International, which contracted his technical services around the world. He trained local youth in any renewable energy technology, from PV to wind to small hydro to solar water heating, even solar cooking. The aid agencies hired his services nonstop, and I would get weekly e-mails from Nicaragua, then Burma, then Mongolia, then the Philippines, then India, then Cambodia, and Peru, Guatemala, Ecuador, Pakistan, Thailand, and Rwanda. I couldn't keep track of him. An e-mail to him during the many years I knew him never went unanswered longer than a few hours, no matter where he was on the planet.

The best was from Rwanda, where he had helped Bob install several multikilowatt PV systems on health clinics sponsored by SELF. He e-mailed me to say, "The two Bills are here." Bill Clinton had been there, and Bill and Melinda Gates arrived shortly thereafter to learn about their foundation's AIDS programs in Africa. He attached a photo of himself in front of a small audience of eager listeners whom he was lecturing on how solar power worked. In the front row, fully attentive, were Bill and Melinda. Bill Gates was astonished to learn that only 2 percent of the people of Rwanda had electricity.

Walt was kind, patient, tireless, and one of the two or three smartest people I'd ever known. He had a quick grin, a quicker wit, and couldn't be angered. He ran his little organization on a shoestring, not concerned about a salary, only expenses. SELF, UNDP, USAID, and other charities got their money's worth when they hired Walt to execute a renewable energy project. He worked for SELF in numerous countries, including Haiti, helping with the electrification of Paul Farmer's rural health clinics. (Paul Farmer, in case you aren't aware of his work, is a modern-day Albert Schweitzer.)

But it was USAID who sent Walt back to Haiti, putting him up in Port-au-Prince at the five-star Hotel Montana, the kind of place he never stayed at. He was in his room when the hotel collapsed on that horrible January day in 2010, when the earthquake struck. It was six weeks before they found his body.

As SELF wrote in its annual report, "Walt's life and work touched tens of thousands of people in developing countries on nearly every continent. There are no words to express the depth of our sorrow at this tremendous loss. Walt, you are, and will always be, an inspiration to us all." The global solar-aid movement had lost its greatest practitioner.

Charitable efforts to light up people's lives with solar electricity and end "energy poverty" around the world continued, with many new players entering the arena. The aforementioned Jeremy Leggett, after launching what was to become the fastest-growing solar company in Britain, also started SolarAid, a nonprofit funded in part with 5 percent of the profits of his new venture, Solar Century (see chapter 12). SolarAid, with backing from corporations and investment bankers, individual donors and funds from aid-agency contracts, created a new business model called SunnyMoney to bring low-cost solar lighting to people in eastern and southern Africa. SunnyMoney franchisees sell "light in a box," a self-contained package with a small solar panel, a long-life battery, and a very bright LED lamp. They call this "microsolar." SolarAid also offers "macrosolar" projects, providing larger lighting systems for schools, clinics, and community centers.

A new method for financing solar home systems is the mobile-money transfer scheme, which uses cell phones and is being pioneered in Kenya by Vodaphone/Safaricom. M-KOPA, a Kenya start-up, is selling $200 electrical systems powered by solar with LEDs and phone-charging kits that can be paid for in small installments through customers' mobile phones.

SolarAid, like SELCO, focuses on commercial distribution of solar, not technology. I've always been skeptical of the "techno fix" and the often well-designed but entirely unreliable solar gadgets that will supposedly solve the unending problem of "two billion people without electric lights." Solar lanterns were the first of these, and while millions have been sold, many were unreliable, giving solar a bad name. However, the electronics have improved, LEDs revolutionized lighting, and now even Coleman makes bright, reliable solar camping lanterns for the U.S. market. Good Indian-made solar lanterns now retail in Kenya for $24.

D.Light, founded by two Stanford Business School grads in 2008, is selling tens of thousands of its Chinese-made portable solar-LED lanterns in Africa and India. The company seeks to reach 50 million people by 2015, replacing their kerosene lanterns. The D.Light is aimed at the very poor who cannot afford solar home systems.

Another social venture, Eight19, spun out of research at Cambridge University, has come up with a similar tiny, low-cost solar-lighting device, which it is seeking to get distributed to very poor people. The goal is to reduce their reliance on costly, dirty kerosene, for which they often spend up to 30 percent of their income. Eight19, a UK-based enterprise, takes its name from the time it takes sunlight to reach the earth—8 minutes, 19 seconds. The group not only wants to address the two-thirds world electricity challenge, but it wants to do it as a fabricator of low-cost third-generation solar cells based on printed plastic using organic carbon instead of silicon. I've seen this story a dozen times over the years, none successful, but one can only applaud their efforts and idealism.

In order to make such high-tech devices affordable, they need to be manufactured "in-country," not imported by charities, says Mark Hankins, Kenya's veteran solar entrepreneur and a friend of twenty years. "Eight19 is another in a long series of companies that installs twenty systems in Kenya on a trial basis, then paints news all over the Internet about their success," Mark told me. He wished them well but said he wasn't sure this was the game changer in solar they claimed it was.

Mark, who is now running African Solar Designs, which installs multikilowatt solar power systems all over East Africa, said, "A cacophony of donors, social entrepreneurs, consultant-planned boondoggles, foundation-sponsored NGOs, poorly planned and executed subsidies, and corrupt government procurement makes it difficult to be a private-sector solar player." Mark grew up in Bethesda, Maryland; married a lovely Masai woman; and is raising two children in their solar-powered house on the slopes of Kenya's Ngong Hills. He was sent to Kenya by the Peace Corps in the 1970s and never left. Despite his jaundiced view of the *cirque du soleil,* he believes "the future is bright" for solar in Africa. "Interest has never been higher in solar energy, and the dawn of a real solar future has come. . . . We can see the day when solar will just be the best choice." It will be delivered commercially and profitably, not by governments or NGOs or donors.

A solar entrepreneur fostering commercial distribution of small, affordable solar in Africa is Doug Vilsack, a young Denver lawyer and the son of the Obama administration's secretary of agriculture, Tom Vilsack, previously governor of Iowa. Doug founded Elephant Energy, and he contacted me in 2008 to say he'd been inspired by SELF and wanted to do something, and so he did: he carried Elephant Energy's first fifty solar flashlights in his backpack to Namibia. I was in awe of his idealistic determination. Three years later, Doug had organized a chain of twelve "energy shops" in Namibia, providing solar lights to 16,000 Namibians. He's aiming to reach as many of the 230,000 off-grid homes in Namibia as he can. His parallel organization, Eagle Energy, provides solar to some of the 18,000 Navajos in the United States who are still without access to electricity.

These "solar lights" are not solar home systems, and they don't provide power for TVs or radios or fans, only very bright illumination. But they are all that some one billion people—out of the oft-referenced "two billion without power"—can afford. A rechargeable, durable solar flashlight, which can run for hours suspended overhead in a home or can be used by women fearful of going outside to the latrine at night, can change a family's life, costing a fraction of what kerosene does and providing ten times the illumination. You have to travel in the unelectrified villages of the developing world to know just how dark dark can be.

One of the breakthroughs in solar flashlights came from a tall ex-marine and former CIA agent from Texas who had served the agency in a dozen very poor countries around the world, including working for Charlie Wilson in Afghanistan (*Charlie Wilson's War*). Mark Bent saw the poverty and the need for lighting, so he resigned from the CIA and designed the world's best solar flashlight with bright LEDs that lasts 100,000 hours and is powered by three rechargeable AAA batteries and a small solar cell. Rechargeable batteries last three years. The flashlight is waterproof, easy to use, and colorful—he makes pink ones for women that the men won't steal. It will sustain 750 to 1,000 charges, or two to three years of night use with a day's charge, before the batteries have to be replaced. The only problem is keeping them from getting stolen when left out in the sun for the day.

I've owned many solar flashlights, and most never worked well, all made badly in China. Mark's are fail-safe. They are also made in China, to keep costs down, but he demands top-quality product: his first run of ten thousand lights didn't meet his standards and had to be destroyed. Not only did he come up with a sturdy, useful product that a billion people could use right now if they could get it, he also came up with an innovative way to finance his commercial enterprise: he sells them in the United States for $29 each or $59 for two, which covers a second light given free to a worthy family in the two-thirds world. He calls this "Buy One Give One," hence the BoGo Light.

I met Mark at a Clinton Global Initiative meeting in New York, and he told me, "You get ten years of use, which benefits ten people on average, and I can sell them to USAID or the UN for ten dollars apiece. Ten, ten, and ten. A pretty good deal for the aid agencies." The Clinton Foundation helped him distribute one hundred thousand lights in Haiti after the earthquake. NGOs like EarthSpark (supported by rock group Linkin Park's Power the World campaign) and Elephant Energy distributed thousands around the world. But Mark can't get the USAID to take him seriously. He told me in 2012 that he flew to meet the USAID director for Afghanistan, where most villages have no electricity, but he got the usual brush-off.

Let me say right here, based on twenty-five years of observation and direct experience, that USAID is the most bloated and useless agency of the U.S. government. The various directors, top political appointees who usually try hard to get things done, normally have no clue as to what is going on beneath them in USAID's vast, unresponsive, and completely ineffectual bureaucracy. It can't be fixed, but we could defund it.

Mark Bent took his lights to the U.S. military. "I told one general in Afghanistan that we should be dropping BoGo lights with little parachutes from our drones, instead of bombs, marked 'Gift from the U.S. people,' and we might win the war. He thought I was nuts," said Mark. Facing increasing quality, corruption, and patent issues in China, Mark wants to move his BoGo lights manufacturing to Houston and employ Iraq and Afghan war veterans, but he needs to find the investment capital. Home Depot has offered to sell his lights, but he needs reliable production volume. Profits would go to the poor of the world without light, the people that Wolfgang Palz targeted twenty years ago, whom solar technology, now enhanced with LEDs, would benefit the most.

Hundreds of small entrepreneurs are now creating their own small solar lighting systems in simple workshops around the world. The German-Swiss Solar Energy Foundation attacked energy poverty in Ethiopia by setting up a small factory to manufacture low-cost solar lighting kits to be sold through franchisees and supported by local microfinance institutions. By 2011, they had installed over 15,000 solar systems, including some in 154 village schools, while employing 105 Ethiopians.

Tanzania has one of the lowest rates of electrification in the world. In Mwanza, the country's second-largest city, Zara Solar is providing high-quality yet affordable solar home systems to thousands of families and institutions on a profitable basis. In Ghana, where only 57 percent of the country is electrified, Deng Solar provides power systems not only to rural people but also as backup to the grid in towns where power cuts average twelve hours a day because there is no rain to supply the main hydropower dams. "The sun is Ghana's oil," says Deng Solar's proprietor. In Laos, Andy Schroeter, a young, idealistic European whom I first met at a Bankgok solar conference fifteen years ago, now runs Sunlabob, a commercial venture in Vientiane, where business is booming. Eighty-five percent of Laotians live in rural areas without electricity.

In Mali, a local, educated youth returned from his good life in Paris to bring solar to all the villages he could reach. Incredibly, to keep costs down, he imports broken silicon solar cells from European PV manufacturers, then recuts them to make small modules, which he laminates in his Bamako workshop. They work! A one-hour video about Afrique Solar was produced by Claire Weingarten, the daughter of Joel Weingarten, my accountant of twenty-five years.

In Guatemala, Juan Rodriguez, a young entrepreneur I met at the Unreasonable Institute in Boulder in 2012, told me about his successful enterprise, which is bringing electric light to thousands of villagers beyond the reach of the grid. His company, Quetsol, imports fully integrated solar lighting systems in four sizes from China, where he'd gone to procure custom-made kits similar to the ones Wang Anhua had built in his workshop in Gansu. His "unreasonable" venture has sold thousands, and was now developing an electronic form of credit payment through cell phones, which everyone has even if they don't have electricity in their homes.

It seems every time I turned around, someone else had joined the *cirque du soleil,* all to good and wonderful purpose. Today, dozens of small NGOs around the world are devoted to bringing solar power and light to the world's poor. And, more amazing to me, nearly twenty years after launching SELCO, hundreds of commercial companies like Juan's are successfully achieving the same purpose, belying the myth also proven wrong by SELCO's Harish Hande that "you can't run a commercial venture while trying to meet social objectives." People are doing it, profitably and happily.

Go to Google or YouTube and you will find thousands of videos about rural solar solutions, technologies, charities, and enterprises worldwide that inspire and amaze. Contact information for many of the examples cited in this book can be found in Appendix 2.

In July 2013, President Obama pledged $7 billion for electrification in Africa over the next five years. "Power Africa" will help, if the U.S. government follows through. However, the International Energy Agency estimates it would take $300 billion to achieve universal access to electricity in the Sub-Sahara by 2030.

From his retirement in Sri Lanka, Anil Cabraal now manages Lighting Africa, a World Bank program "helping to develop modern off-grid lighting markets in Sub-Saharan Africa." These are commercially available stand-alone solar lighting kits that can be installed by a typical user without employing a technician. Lighting Africa's goal is to reach 250 million people by 2030. Anil, the father of most of the renewable energy programs at the World Bank, which today still devotes only 1 percent of its energy lending to solar, is confident solar entrepreneurship "catalyzed" by bank funds is the way to go. "Africa needs solar entrepreneurs that can convince buyers to go solar, instead of buying generators, which are marketed as 'classy,'" said Mark Hankins. Mark, like Anil, believes in Africa's enormous untapped solar markets, from tiny lighting kits to grid-connected PV systems. His company right now can't keep up with demand. In 2010, Mark Hankins took time off from developing solar projects all over Africa to write the finest book available on off-grid PV: *Stand Alone Solar Electric Systems: The Earthscan Expert Handbook for Planning, Design and Installation*.

A dozen years ago, and a decade after I left my media-director post at Greenpeace to start SELF, Greenpeace at last launched a global solar campaign. In a paper outlining a "strategy to deliver renewable energy to the world's poor," a Greenpeace campaigner wrote, "As we begin the 21st century, one-third of the world's population are still living without access to electric lighting, or adequate cooking facilities. Providing these two billion people with modern energy systems . . . is *one of the most pressing problems facing humanity today*" (italics in original). Noting that the G8 had set up a renewable energy task force "calling for 800 million people in developing countries to be provided with energy from renewables within ten years," Greenpeace pointed out that this would still leave "over one billion people dependant on unsustainable energy sources, resulting in severe environmental and social impacts." Interestingly, the chairman of the G8 renewable energy task force (on which I served as an advisor) was the retired chairman of Royal Dutch Shell, Mark Moody-Stuart. In our occasional e-mail exchanges, I never summoned the courage to ask him why—why him? How did he go from heading up the world's second-largest oil company to chairing an international forum on renewable energy?

Back in June 2004, former Soviet leader Mikhail Gorbachev announced what would become an unfulfilled commitment to "fight energy poverty for two billion people without electricity" and called on the World Renewable Energy Conference in 2004 to adopt his plan for a $50 billion "global solar fund." Gorbachev's Green Cross organization, which had conferred the Millennium Award for International Environmental Leadership on SELF in 1998, pointed out that such a global financial commitment to solar power would lower the cost of photovoltaics, address "energy poverty," deal with urban peak demand (which caused the northeastern blackout in 2004), and create an "energy glasnost" to open electrical grids to flexible, decentralized, smart energy solutions. He was way ahead of his time. As you can see, once again, when it comes to solar power for the world, there is nothing new under the sun.

SELF continued to earn awards too numerous to mention. In 2008 Bob and I were both invited to the Aspen Environmental Forum, where I spoke about SELCO, and Bob received the King Hussein Environmental Leadership Award personally from Queen Noor. Bob introduced me to the gloriously beautiful and intelligent former queen of Jordan and later thanked me from the stage for entrusting him with SELF since 1997. I'd never called anyone "your

SELF Exec. Dir. Bob Freling accepting King Hussein Environmental Leadership Award from Queen Noor at Aspen Environment Forum, Colorado. *(Author)*

highness" before, and I talked with Queen Noor about skiing and solar, and I found myself in a group of onlookers, where I happened to introduce her to the former head of the CIA, James Woolsey, now a big solar promoter, who said, "I knew your husband." Of course! I felt like an idiot, although I'm not sure the queen knew who Jim was, so I guess I'd done my social duty.

In 2012 the United Nations sponsored World Sustainable Energy Year (which coincided with the twentieth anniversary of the Rio Earth Summit). The goal is universal energy access for the 1.5 billion people (the UN's number) who lack electricity. Echoing Gorbachev's proposal seven years earlier, an International Energy Agency (IEA) study in 2011 said that $48 billion a year could bring electricity to one billion people in twenty years using "low carbon energy." Said the IEA executive director, Maria van der Hoeven, "Eradicating energy poverty is a moral imperative and this report shows it's achievable." Bob Freling's term "energy poverty" had gone mainstream, but his "energy is a human right" motto remained a clouded concept as long as the World Bank, the United States, and the EU—despite the heroic efforts of Wolfgang Palz—failed to devote the necessary resources to close the energy divide. Republicans claimed we were wasting money on foreign aid, when in fact we spent a comparative pittance. Instead, America chose to fund the military, accounting for more than 40 percent of the money spent on defense worldwide.

Unfortunately, population growth—*the planet's biggest challenge*—adds people faster than all the governments, utilities, aid agencies, NGOs, and businesses can increase the power supply to meet the growing needs of. These people may live out their lifetimes without ever seeing electric power like that enjoyed by the rich nations since Edison's time.

The Solar Electric Light Company, meanwhile, had prospered in the difficult retail markets it had chosen to focus on in India, Sri Lanka, and Vietnam. SELCO's 320 workers and managers had a never-say-die, can-do attitude. The future belonged to them, if not to the company itself. But since this was capitalism, not philanthropy or socialism, the company belonged to the shareholders, not the workers, and we had to follow capitalism's rules, including the golden rule—those with the gold rule. Most of our shareholders were exceedingly patient; they had invested in SELCO knowing full well that this was a long-term proposition requiring more than usual patience. Both Rolf Gerling and Stephan Schmidheiny were patient (billionaires can afford to be), willing to risk a few million dollars to put their money where their mouths were, in support of sustainable development.

Two weeks after we raised our initial capital in the go-go 1990s, the Asian financial crisis hit—heralding the bursting of future bubbles, like the dot-coms, that would go the way of Bangkok real estate—but it had little effect on the company. SELCO survived the implosion of the stock market in 2001, which left millions of Internet-company investors holding worthless stock, and the corporate scandals that left thousands of shareholders with nothing. SELCO survived Enron and its U.S. debacle, and SELCO-India survived Enron's $2 billion disaster at Dabhol and the new globalization that flooded India with cheap goods and commodities from China. SELCO-India is our biggest success story, making its first audited profit in 2002. "I can't believe we're actually making money selling solar energy to poor people," Harish Hande told me. The profit was slim, and a long time coming, but it was there. The parent company finally reached breakeven in 2004. However, SELCO could never figure out how to repatriate funds from its subsidiaries and to recoup investment capital in some form. We never reached the hugely ambitious targets we'd projected in our business plans, despite high annual growth rates.

I stepped down as CEO in 2002, for I believed it was time to find a younger leader and to begin the process of pushing the overall managerial functions down to country-operation levels so we could close the expensive Maryland headquarters—SELCO's key to success was trusting local foreign nationals to do the job. I finally left the board and resigned as a director in 2004, after new investors bought out our largest shareholders and took control of the company. Their deep pockets suggested that SELCO would not be without financial backing for growth and expansion in the future. Harish, a natural leader, eventually stepped into the top position of the restructured company.

By 2013, SELCO-India had sold and installed over 150,000 solar-home-lighting systems. No other company had achieved this. It was selling energy services to underserved populations, and its revenues were growing at 35 percent a year. It was neither extracting resources nor exploiting workers, as happens too often with foreign direct investment. SELCO sought to reduce energy poverty among people in need, not exploit energy poverty as Enron had tried to do in India. We had proved the market economy could serve people first, and by serving people we built a successful company. We had learned many lessons along the way.

In his two books, *Financing Change* and *Changing Course,* which are still used in many U.S. business and economics courses, Stephan Schmidheiny, our original investor, has written that companies should not have to suffer from the tyranny of the quarterly statement imposed by capital markets that demand growth at all costs. This is certainly true for a private company like SELCO, which had no need to impose anything but annual targets based on realistic expectations. New values must apply if "sustainability" is to be possible.

SELCO's actual business on the ground in the rural areas of the developing world could never be reconciled with the values of Wall Street, just as the world of finance capitalism can never be reconciled with the market economy, free enterprise, fair trade, and sustainability. As David Korten had written in *The Post-Corporate World*, "Finance capitalism makes money from money, without the intervening necessity of engaging in productive activity. . . . In a market economy, investment is about creating and renewing productive capacity to meet future needs. In a capitalist economy, investment is about making money." Those words were written a decade before Wall Street's criminal shenanigans and greedy bankers brought down the global economy in 2008.

Our directors and shareholders' representatives (but not necessarily the shareholders themselves) could not understand that the foundation of our business was *service*. They could not see that the SELCO companies existed *only* because our hundreds of workers were dedicated to the company's *mission* and often worked without pay when the revenues could not meet the payroll. Mission and service came before fear and greed.

When auditors and financial analysts derided our little company for its slow growth, financial inexperience, managerial deficiencies, and what seemed to them to be an irresponsible, steadfast pursuit of an impossible dream, I would look at the photo on my bulletin board of Enron's Ken Lay, that beacon of corporate savvy, rectitude, and experience, the king of big-growth capitalism—and personal friend of President George W. Bush—being led away in handcuffs. It made my day.

We had no experience as businesspeople or energy entrepreneurs, only some vague idea that solar power would be good for poor people in developing countries who had no electricity and no prospect of ever getting any unless a commercial enterprise could deliver it to them. Only fools rush in to the most difficult and frustrating energy market in the world, notwithstanding that it's also the largest. But we were happy fools because we all felt very good about how we were seeking to make a living.

I decided to tell our story in a book, *Chasing the Sun,* published by a small Canadian publisher in 2005. An associate managed to get a copy to President Clinton, knowing of his keen interest in solar energy. A couple of months later my phone rang, and it was the Clinton Foundation inviting me to attended, gratis—waiving the $25,000 registration fee—the next Clinton Global Initiative (CGI) meeting in New York. "Bill read your book. He loved it," said the voice on the phone. "He's even ordered a bunch for the Clinton library bookstore." I began to doubt the former president's critical faculties: my book certainly wasn't *that* good.

Only President Clinton could organize anything as remarkable as the CGI, inviting over one hundred heads of state—in town for UN Week—and the globe's leading movers and shakers. At the 2006 CGI, I found myself sitting in the Energy and Climate Change workshop with Richard Branson; Warren Buffett; Shimon Peres; Barbra Streisand; Wesley Clark; John Glenn; billionaire Vinod Khosla; Chris Flavin, of Worldwatch; Frank Tugwell, of Winrock International; *The New York Times*'s Thomas Friedman; Amory Lovins; Jeremy Leggett; and the late Wangari Maathai, whom I had previously met in Kenya. Present were the CEOs of Sony, Cisco, Wal-Mart, Swiss Re, and India's Infosys, among others. What was I doing here? I never knew, but I kept getting invited back because renewable energy for the developing world was an ongoing theme of CGI during the next four annual meetings I attended.

At the first one in 2006, my mind numb from the speeches by Al Gore, Bill Gates, Mohammed Yunus, Tom Brokaw, Rupert Murdoch, Colin Powell, Desmond Tutu, Kofi Annan, Hamid Karzai, Queen Rania of Jordan, and many others, and after chatting briefly in the hallway with Sergey Brin and Larry Page, founders of Google, about their new multimegawatt solar installation on their corporate headquarters, my cell phone vibrated. It was Bob Freling, also an invitee (as was Harish Hande in later years), telling me to my utter amazement, "President Clinton wants to meet you. I'll see you downstairs."

We entered the Sheraton New York's huge ballroom, where Clinton had just moderated a panel and was now surrounded by hundreds of admirers from around the world, and Bob wedged his way through the crowd, straight up to the president, and said, "I'd like to introduce Neville Williams." (Bob had not only met Bill in Rwanda, but he had been invited to the first CGI, where SELF provided the carbon offsets that had been calculated to render the meeting "carbon neutral.")

"I'm so glad to meet you. I really loved your book. Thank you so much for sending it to me." He grabbed my hand in both of his, and continued effusively. "I was so excited about that book. I was like a high school kid, sharing it with my friends—I told Hillary and Chelsea to read it. It's a great story. You've done really good work."

I was so dumbstruck I couldn't remember to tell him I loved *his* book, the recently published autobiography, *My Life*. He said he had put solar on his presidential library and planned to install it on his home in Chappaqua. I'd never met a president before and remained speechless. I could only get out a "thank you for inviting me" before the Secret Service led him away as crowds continued to press in.

The best thing about CGI was seeing world leaders and CEOs focused for three days, at each CGI meeting, on clean energy solutions to poverty and climate change. Al Gore denounced "clean coal" as a fraud and said people should occupy coal plants to shut them down! Clinton himself talked extensively about solar photovoltaics, in particular. It was too bad the Clinton-Gore administration had not done more to combat global warming and fossil fuels, but now they were making up for it, and it was indeed inspiring to know that the most powerful people on the planet supported "clean energy now," the theme of Earth Day 2000. People were getting the message, at last. Warren Buffett, who was sitting at the next table in our energy breakout room, got the message and in 2012 bought a $2 billion solar farm in California.

In Germany, Hermann Scheer liked my book, too. We'd known each other for nearly fifteen years, since our first meeting in Zimbabwe. Besides serving in the Bundestag, he had continued writing books, and I'd read them all. He wrote them in longhand on yellow pads, in German, in a powerful, authoritative, and aggressive style. He also had time to be the president of EUROSOLAR, the EU lobbying and educational organization he'd founded with his wife in Bonn.

In Berlin one day in 2000, he asked me and my friend Michael Eckhart to set up the American Council on Renewable Energy (ACORE) to partner with his World Council on Renewable Energy. I declined to volunteer for this, but Mike jumped at the chance, and he would later turn ACORE into America's largest renewable energy organization. Scheer also set up the International Renewable Energy Agency (IRENA) to take the renewable energy message to world leaders and the UN and to counter the International Energy Agency (IEA), which was committed to nuclear and fossil energy.

Scheer told the Berlin audience, always in his lilting, forceful, German-accented English, "The transition to a hundred percent renewable power supply represents the most comprehensive economic restructuring since the beginning of the Industrial Age, and it is inconceivable to imagine that the process will not have both winners and losers." He explained how rigid, unimaginative, conventional thinking impeded change. "For the traditional power industry, a system of widespread, decentralized production based on renewable energy is almost inconceivable. Such structurally bound ways of thinking explains how IBM missed the PC revolution, believing the future of computers lay in centralized processors."

At Mike's landmark ACORE conference at the Capitol's Cannon Caucus Room in 2004, Scheer, the keynote speaker, further explained how we could "break the globalized chains of the fossil-fuel supply networks," which have tied down the global economy for so long, running it like a private cartel.

While he had relentlessly pushed the EU to fund solar for developing countries, his reputation as the "father" of the German national solar photovoltaics program had been gained, after he pushed through the Renewable Energy Act of 2000, which bypassed government, industry, and the vaunted "market" to institutionalize the use of PV *by the people* . . . who voted to buy it, to use it, to pay for it, and to guarantee its growth as a mainstream energy business.

"We didn't wait for the market or for a consensus or for the conventional energy system to do the job," he told me. "There is no free market in energy, and governments are the last to move. So, we relied on the people. We went to the people with the initiative, who voted in favor of solar electricity. But we are in a race against time to replace conventional energy with renewables. What we need now are practical protagonists" (see chapter 12 for more on the German solar program).

The "practical protagonists" remark stuck in my head, as you will see.

A year later, after Hermann read my book, his office called from Berlin to invite me to the World Renewable Energy Assembly in Bonn, where he was to be given the first Einstein Award by Germany's largest solar company, SolarWorld (which had bought up the assets of Siemens Solar worldwide). I was asked to be the presenter and said I'd be proud to do so. So at a formal gala at the Museum of Natural History in Bonn on February 27, 2005, after greeting the crowd of German politicians and their spouses with "*Guten Abend, Meine Damen und Herren,*" then switching to English, I honored Germany's "solar king" as best I could. There was not much I could say that wasn't already known about him, but I added that, ironically, because of the enormous success of the German PV program, the "power for the world" efforts that he and Wolfgang Palz championed had been negatively affected. It was now almost impossible for companies like SELCO in the developing world to get 40- and 50-watt solar panels, since PV production worldwide was all going to Germany, and the German market required panels of at least 200 or more watts. Prices of small panels, if you could get them at all, nearly doubled! (This setback would later be rectified as more PV manufacturers entered global markets to serve all levels of demand—there are some fifteen hundred today.)

Frank Asbeck, Germany's leading solar entrepreneur and the CEO of SolarWorld AG, had said in his introductory remarks, "The life and work of Albert Einstein are a symbol of the unity of scientific and social commitment. It is this exemplary unity of science and ethics that we want to honor with the SolarWorld Einstein Award." Following my own remarks in praise of my friend, he handed me the prize to give to Hermann, a small bronze globe, a plaque, and a five-mark silver coin minted with the image of Albert Einstein, the first father of photovoltaics. It was a much bigger honor for me to present this award than it was, I'm sure, for Hermann to receive it. In any case, just as Einstein had won the Nobel Prize for his discovery of the photoelectric effect, I believed Scheer deserved a Nobel Prize for his lifelong political advocacy of solar power worldwide and for the effective implementation in Germany of his vision of the solar age.

The next day, Frank Asbeck invited me to his home, a fifteenth-century castle on the Rhine. It was like nothing I had ever seen. His children romped and servants decorated the great room with Christmas boughs as a fire roared in the massive stone fireplace, and paintings of his Prussian ancestors stared down from the walls. He'd bought Bonn's finest residence with proceeds from his booming solar company, which made him Germany's first "solar billionaire." We talked about the solar business, and I told him now that I was done with SELF and SELCO and had written my book on solar, I was thinking about how to sell solar to Americans. I told him I was going to start a new company, this time in America. What Germany, Hermann, and SolarWorld had achieved in a very short time had inspired me.

I gave him our business plan for Standard Solar Inc.

11

At this point, the obvious question remains: If millions of poor families in the two-thirds world can get their electricity from the sun, why can't we? Or, if one small company like SELCO, plus hundreds of smaller players around the world, can deliver solar power and light as an affordable option to families, businesses, and institutions in some of the least economically advanced countries, what is so hard about doing it in North America and Europe?

I started pondering this question after I submitted the manuscript for my first book on solar to my Canadian publisher. Researching the final chapter, I had investigated solar developments in the United States and Europe and had written about the PV industry and solar entrepreneurs. It had been twenty-five years since our Gold Lake energy conference in Colorado, where Tom Tatum and I tried to figure out how to sell solar to the American people, and failed. This was 2004 and the Iraq war was raging, the dollar was falling, China was booming, and OPEC was scheming. Tom called me up, Cassandra that he is, to say, "Fifty-dollar-a-barrel oil by fall." I didn't believe him. It hit $50 the first week of September and $55 in October. Wouldn't we love to go back to $50-a-barrel oil today!

The cost of oil, of course, doesn't directly affect electricity prices; it affects transportation. But since everyone in North America drives, everyone knew "energy" was getting more expensive. Books appeared in 2004, such as *The End of Oil*, by Paul Roberts; *The Party's Over*, by Richard Heinberg; and *Out of Gas*, by the Cal Tech professor David Goodstein. And there was lots of talk about "peak oil." Oil production "peaks" when you extract more oil than ongoing exploration can discover in a particular market or country or oil field. U.S. oil production peaked in 1970, and in many other oil-producing countries it peaked not long after that. In 2004 it looked as if oil was going to peak globally much sooner than anyone realized. Roberts wrote, "Oil depletion is arguably the most serious crisis ever to face industrial society." Professor Goodstein, in his devastating history of fossil fuels, wrote, "Our way of life, firmly rooted in the myth of an endless supply of cheap oil, is about to come to an end. . . . We will for the first time in history be consuming oil faster than we are finding it." He concluded that *if we don't turn to the sun for energy,* "civilization as we know it will come to an end sometime in this century unless we can find a way to live without fossil fuels." Solar power will enable us to win "the real war," he wrote. "The alternative is to go on hunting terrorists while our civilization slides into oblivion." (We're still finding oil, and today natural gas is abundant and cheap—see chapter 12.)

In 2004, the fiftieth anniversary of the invention of the solar cell was celebrated. It was also a watershed year for other reasons. Four destructive hurricanes in Florida knocked out power to millions. The northeastern blackout plunged 50 million people into darkness as overloaded power lines struggled to meet demand during the hot summer. It was also the fourth hottest year on record (nine of the ten hottest have occurred since 1995). That year boasted the warmest October ever and a record number of tornadoes (1,734) in the United States, since surpassed in later years. It was the costliest year (so far) for the insurance industry, which faced $36 billion in claims, largely due to the greatest occurrence of weather-related disasters ever in a single year. (Those terrible weather events were to be eclipsed tenfold in the coming decade. Witness the hurricanes Katrina, Rita, Ivan, Irene; the destruction of Greensburg, Kansas, and Tuscaloosa, Alabama; and the first "superstorm," Sandy.) The hottest year ever, so far, was 2012, during which extreme weather caused $110 billion in damages.

These things get on people's nerves.

Three things were happening simultaneously: (1) people became aware of climate change caused largely by two centuries of greenhouse-gas emissions, and many wanted to do something about it; (2) energy prices—electricity rates—were rising everywhere; (3) Americans saw "energy security" as an issue as the United States continued to import more than half its oil from abroad while going to war in the Middle East to protect oil supplies, calling it a war for freedom and democracy. These three unfolding "events," of which we all became acutely aware in mid-decade, had little to do with one other but everything to do with solar and the other renewable energies.

In 2004, when I would tell people about our work providing solar power to poor people around the world, the question invariably was, "What about us? How can I get a solar system for *my* house?" I really didn't know, other than to advise them to call one of the several "mom and pop" solar installers operating in the Washington area.

That's when my wife said, "Why don't *you* start a solar company *here*?"

At first I said, What? Me? I just got out of running several companies and organizations with all the responsibility that entailed. I'd written a book. I looked forward to more writing, having forgotten that you can't make a living as a writer. Besides, I was sixty-two. But then I thought, well, I don't really have anything else to do. I reread a paragraph in the last chapter of my forthcoming book that asked, "If illiterate farmers could figure out that solar is better for them than the alternatives—kerosene and darkness—why couldn't we make a similar life-changing decision to install an affordable clean energy system in our homes?" I had written, "If the solar age is to be about you and me, we need to be the customers for solar power."

If American families were to "go solar" they first needed to know who to call! There had to be local, professional installation companies just as there were in India. So I took my wife's advice and decided to start one. But I had no money, since my common shares in SELCO became worthless when the preferred shareholders restructured the company and the common shareholders were "crammed down." The advance for my forthcoming book was miniscule. While I tried to figure the next step, I continued researching the U.S. solar industry circa 2004.

In December of that year I attended the third conference of the American Council on Renewable Energy (ACORE), which is part of the World Conference on Renewable Energy (WCRE) that my friend Mike Eckhart had founded at the behest of the late Hermann Scheer. The five hundred attendees in regulation dark suits sat attentively in the magnificent Cannon Caucus Room of the U.S. House of Representatives on Capitol Hill and listened to speeches about how solar was going mainstream. Three years earlier, Senate majority leader Trent Lott had publicly dismissed solar power as "hippie technology." These weren't hippies; they were all middle-aged businessmen and policy makers.

Admiral James Woolsey, former director of the CIA, addressed the gathering, saying, "If you drive an SUV, you may be contributing to the sinking of Bangladesh." He drove a Prius and had just put both a solarelectric and a solar hot-water system on his house. Woolsey said we had no choice but to use alternative energy sources because we were going to be at war with militant Islam for decades—and "it is a war we may not be able to win." It was clear he and fellow speaker Robert "Bud" McFarlane, President Ronald Reagan's national security advisor, were concerned that Bush's invasion of Iraq was not helping win this war but instead had unleashed the whirlwind. McFarlane, one of the enablers of Reagan's infamous Iran-Contra episode, also drove a Prius and said we had to get off foreign oil. Security expert Frank Gaffney implied that "Islamic fascism" was a worse threat than Nazism or communism ever were, so we'd better damn well get serious about finding alternatives to the oil that Islamists control. These were all conservative Reaganites at a solar conference talking about global warming and energy security!

The president of GE Energy, the world's largest manufacturer of wind turbines, announced that "GE is betting on a greener future." That year, GE purchased the assets of Astropower in Newark, Delaware, once the largest American-owned manufacturer of solar photovoltaics, which GE planned to turn into a billion-dollar business.

"Edison would be pleased," I told him afterward.

"He's our founder," he said proudly.

"I know," I said, reminding him of Edison's quote in favor of using the sun's energy: "What a source of power!"

GE got out of solar panel manufacturing nine years later, but remained active in solar power generation—see chapter 12.

While we inelegantly munched our brown bag lunches catered by the Republican Club of Capitol Hill, Amory Lovins summarized his new 2004 study, commissioned by the Pentagon, "Winning the Oil Endgame." Winning that game, he asserted, will be the "cornerstone of the next industrial revolution . . . and, surprisingly, it will cost less to *displace* all of the oil that the U.S. now uses than it will cost to *buy* that oil." Ever the dreamer, Lovins nonetheless was right, about as right as President Carter was in 1979 when he had said, "We must end our dependence on foreign oil, and we must get twenty percent of our energy from the sun by the end of the century." He was speaking of the *last* century. In 2012 Lovins would release his brilliant blueprint for a new energy economy in his book, *Reinventing Fire,* asserting that we will definitely replace coal and oil with renewable energy *this* century. We will have no choice.

Historical perspective at the ACORE conference was provided by Jay Hakes, director of the Jimmy Carter Library and Museum, who was previously the director of the U.S. Energy Information Agency; he showed us an archival video of one of Carter's visionary solar speeches of twenty-five years earlier, astonishing the younger members of the audience who thought solar was something new. Hakes reminded us of what President Richard Nixon said in 1973: "Let us set as our national goal, in the spirit of *Apollo,* with the determination of the Manhattan Project, that by the end of this decade we will have developed the potential to meet our own energy needs without depending on any foreign energy source." Amory Lovins and Richard Nixon as coequal visionaries! That evening, Jay told me over dinner how much Carter, trained as a nuclear engineer, truly believed in an America powered by solar energy one day, and I told him how Tom Tatum and I prematurely shared that vision during our days at the DOE promoting solar for Carter's first assistant secretary for renewable energy, Omi Walden.

* * *

During a break in the conference, I went outside in the crisp winter sunshine and smoked a cigar and thought about what I was hearing in the great Cannon Caucus Room. Something had happened while I was running around the world during the past fifteen years, since I had first consulted for Solarex, which was now BP Solar, the main sponsor of the first Solar Decathlon, held on the Washington Mall in 2002, a wonderful event that had also made my head spin.

That year, 2004, *Boiling Point* was published, written by Pulitzer Prize–winning author Ross Gelbspan. If you are a fan of horror movies, forget them and just read this book; if you don't like to be scared, skip it. Besides reminding us how serious a threat climate change is to our species, Gelbspan took issue with nearly everyone, as evinced in the book's subtitle: "How Politicians, Big Oil and Coal, Journalists, and Activists Have Fueled the Climate Crisis—and What We Can Do to Avert Disaster." He castigated the Washington energy- and environmental-policy community, which he said had been hijacked by the fossil-fuel lobby. He offered a Marshall Plan approach to "rewiring the world with clean energy." Like Scheer, he proposed a new industrial revolution . . . nothing short of an "energy revolution."

Also in 2004, that tipping-point year:

* Bill Clinton, in a landmark speech at New York University, said that we should stop "bellyaching and whining" about political obstacles and undertake a new effort to address the "intertwined problems of energy dependence and global warming."
* Bob Stempel, former chairman of General Motors, went to work for ECD Ovonics, manufacturers of Uni-Solar PV modules. He told *Fortune* magazine that "solar power . . . is becoming mainstream . . . growing at 25 percent a year [and] the business case for solar is becoming clearer. . . . Solar can become a free source of power for you."
* In 2004, Americans bought eighty thousand hybrid vehicles while the sales of GM's gas hog Hummer collapsed. (Prius sales would top two million by 2012.)

* The touted "hydrogen economy" was finally seen as the public-relations and financial-investment scam that it was. At the 2004 ACORE meeting, not one speaker even mentioned hydrogen or fuel cells. Solar and wind were now seen as the main new energy sources.
* Fifty-three percent of Coloradans voted for a state Renewable Energy Portfolio Standard, which required its major utilities to buy or produce 10 percent of their power from renewable energy sources by 2015 while establishing net metering standards for grid-connected household PV systems that would become policy in forty-three states.
* In December 2004, California's governor, Arnold Schwarzenegger, proposed a statewide subsidy for solar PV systems, expanding the successful solar incentives pioneered by David Freeman at LADWP. The governor's energy office announced that California could have one million homes and buildings powered by the sun by 2018.
* New Jersey, through its generous Solar Renewable Energy Certificates (SREC) program, created a boom in solar-PV installations backed by the New Jersey Board of Public Utilities with a

Six-megawatt solar farm powers the Lawrenceville School in Princeton, New Jersey. (*Solar World and Lawrenceville School*)

fund of $745 million for solar and energy efficiency. New Jersey would surpass California in new PV capacity during the first quarter of 2012.

* In Washington, D.C., the city council voted to require D.C. utilities to obtain 10 percent of their electricity from renewable energy by 2022. "There's plenty of sun in D.C.," said Rhone Resch, president of the U.S. Solar Energy Industries Association (SEIA). "Every rooftop in the city has the potential to generate its own electricity." I couldn't have imagined it at the time, but two years later Rhone would become my company's first customer when he bought a 7-kilowatt PV system for his home in northwest D.C.

It became clear to me in 2004 that the environmental, energy, economic, psychological, cultural, and informational climates—as well as the *actual* climate—had changed as they never had before. Energy costs, energy security, and clean energy were now popular mantras as literally billions of people became concerned about at least one of these issues, if not all of them. Change doesn't come about because of policy studies, books, reports, articles, strategy meetings, and conferences, although these help; change comes because people want it. When the people lead, politicians follow.

My wife's words were still ringing in my ears, but I was still not convinced I could do anything. Then, I read environmentalist author Bill McKibben's article in the November 2004 issue of *Mother Jones* magazine, titled "One Roof at a Time." McKibben had been a solar skeptic until he put twelve PV modules on his own house and quickly became a believer. He interviewed happy PV users around the country, who said they *loved* watching their electric meters run backward on their grid-tied systems as the solar power they produced fed into the electrical grid when the sun shone. Users of sun power became energy-efficiency enthusiasts so they could make their meters run backward *faster*. McKibben noted that most months his electric bill got "substantially smaller" as he earned credit from the utility. He reported that "more solar power has been harnessed on the world's rooftops in the last two years than in all of previous history."

This was all I needed. I was in. I liked the idea of "saving the world one rooftop at a time." I would do as my wife suggested and start a company. Americans, unlike the poor around the world, didn't *need* solar power, but it was clear that a great many *wanted* it. They wanted it so they could do something about global warming and climate change, and also save money in the future. There was a pent-up demand for solar, nowhere more so than in the greater Washington, D.C., area with its highly educated, upper-income population (89 percent of the residents of Montgomery County had college degrees, and the five counties surrounding D.C. had the highest household incomes in America). Where there wasn't a demand, I believed it would be possible to create one, like Detroit created a demand for the SUV, a vehicle no one ever asked for. I firmly believe in the first capitalist tenet, "Supply creates its own demand."

Many Americans wanted to buy solar *but they didn't know who to call.*

It was time to sell residential solar power to Americans. "Build it and they will come," I hoped. It was time to move beyond the image of "Chuck with a truck" and the ponytailed backyard solar practitioners who dominated the retail landscape thus far. It was time to sell solar in the same way we had sold it in India. To thousands, and profitably.

It is not easy to start a company when you have no money. It means you have to raise it. To do that, you need a good business plan, which normally takes about nine months to write. That means research and information gathering and a lot of work. The first thing I did was take two local solar installers to lunch, each a pioneer in his own right, entrepreneurs that had been meeting local demand as best they could with no capital and few employees. Aurora Energy, which operated out of the proprietor's living room, had installed the solar system on Jim Woolsey's house and put the solar PV panels on the White House maintenance shed. Chesapeake Solar had started in the solar hot-water business and had gravitated to solar PV as more and more people asked for it. I asked both practitioners if either of them wanted to join with me to start a new company, but they were happy with their mom-and-pop enterprises; they encouraged me to go ahead and didn't fear competition, least of all from me. Except for a couple of other "old solar hippies," they were the only game in town. This was not California.

The next thing was to come up with a name. Recalling the importance of Standard Oil as it sought to "light up the world" with its kerosene lamps via a worldwide house-by-house lamp-oil distribution system, becoming the largest corporation in the world, I decided, shamelessly, to call my company *Standard Solar*.

Until now, most residential solar installations were custom, one-off jobs. We decided to standardize them and sell five or six standard solar home systems at an advertised price.

We would sell solar as a consumer product, house by house, one rooftop at a time. I registered Standard Solar Inc. as a Delaware corporation on December 28, 2004, and secured a URL, and later trademarked the name. Now what? I was now president, chairman, and CEO of a solar company with no capital, no employees, and no office.

I needed help. You cannot start a company alone. You need partners, associates, colleagues, friends—and investors. Serendipity was about to kick in. I was told by Bob Shaw, one of the early venture investors in solar manufacturing in the United States, to look up solar veteran Gerry Braun, whom he knew was interested in the prospects for residential solar. Gerry had worked for PG&E in California, running its energy lab, then for the DOE, then had spent twelve years at Solarex, until shortly after it became BP Solar. I recalled having once met him there. He was my age and temporarily out of work. I had no idea where in the United States he presently lived. Serendipitously, it turned out he lived two miles up the road from me in Gaithersburg! We met, and he liked my idea and agreed to help with the business plan as a consultant. I offered him some founders' stock in Standard Solar and said I would pay him for his time when I could. We went to work in my basement office, where Standard Solar was launched.

Gerry was an MIT engineer who knew as much about solar PV as anyone I had met, except maybe for Paul Maycock and Peter Varadi, who also encouraged me to start the new company. The first question I asked Gerry was about competing technologies; should I focus on PV? There was so much talk about fuel cells at the time, which many "experts" had claimed would soon be powering American homes. He told me, "When we researched fuel cells at PG&E in the 1980s, fuel cells were about five years behind PV. Today, they are about fifteen years behind PV." That's all I needed to hear. (I still don't know of a home anywhere powered by a fuel cell, despite the predictions by futurists and the millions invested in the problematic technology—however, a number of businesses use industrial-size fuel cells now [see chapter 12].) Without his help, I couldn't have created a credible business plan, but Gerry was not an installer or businessman, nor was he an entrepreneur.

I needed someone with hands-on experience. I heard about an installer in our county named Lee Bristol, who had just started offering "renewable energy services" out of his home office. I called him up and said I wanted to meet him. It turned he lived only a few miles north of Gerry Braun. When I arrived at his rural home with sweeping views of northern Montgomery County's bucolic farmland, cows and horses everywhere, I noticed a line of solar modules along the ridge top of his south-facing roof.

Lee greeted me warmly and introduced me to his wife and daughter, and we settled into his living room to talk solar. "I heard about you," he said, "and SELF and your work around the world. I was planning to call you one of these days."

"Looks like I beat you to it," I said.

We shared travel stories about Africa, talked about geodesic domes and Steve Baer's Zomeworks, about the Whole Earth Catalog and his hero Stewart Brand. Lee was also an MIT engineering graduate and had opened the first computer-software store in Washington, D.C., in the seventies. Now, he was in charge of IT for Kaiser-Permanente, which paid him well, but the job bored him. "I always wanted to get into solar," he said. "So I took a correspondence course from Solar Energy International in Colorado." I told him I'd been through the same course fifteen years before, in Carbondale, but I wasn't very adept at actually installing solar systems. I could understand solar, but I couldn't "do" solar.

"Did you put up that solar system on your roof yourself?" I inquired.

He had, learning as he went, and it worked fine. It had no batteries but was connected to the power company through an Austrian-made inverter. Lee was among the first in Maryland to take advantage of the new net-metering regulations that allowed small solar power producers to sell power to the local utility in the daytime while drawing the electricity they needed at night from the grid. I watched the disc on his electric meter run backward, very rapidly, under bright sun.

Neighbors in the area were asking Lee to install solar systems for them, since there was no one else nearby to do it: Aurora Energy was in Anne Arundel County and Chesapeake Solar was based in Howard County. Lee was the only "solar guy" in Montgomery County. He'd done a dozen or so installations through his sole proprietorship, LBA Renewable Energy. He even had a Web site.

But he wouldn't quit his day job to take the risk of running a solar company full-time. He had a family to support, and, while he was exceedingly entrepreneurial and inventive, he wasn't a risk taker. That was my job, as always, fool that I was. He agreed, however, to join my board. Now, I had Gerry and Lee, and they got along very well, both MIT grads with linear engineering minds that could focus my fuzzy, abstract, right-brained thinking.

Meanwhile, as someone who has never managed to balance his own checkbook, I needed a financial person, maybe even an MBA. Or at least a good accountant. (I had already learned the hard way that most MBAs know little about accounting, a specialty I had been forced to become familiar with as CEO of my previous ventures.) My friend Scott Sklar, former senator Jacob Javits's chief of staff, who later founded the Solar Lobby and, after that, the Solar Energy Industries Association (SEIA), called me one day to say he knew someone I might want to meet. He said the gentleman was looking to get back into the solar game now that it was getting hot.

I met Tony Clifford at the Hunters Inn for lunch in nearby Potomac, and our Irish temperaments clicked. Yes, he had an MBA; in the seventies he'd graduated from the University of Virginia Darden

School of Business and went straight to work for my friend . . . *Peter Varadi* (and Joseph Lindmyer), who had just founded Solarex at the time. He was their first "MBA hire." Serendipity had clicked in once again.

Tony had been out of solar for the past twenty-five years and was currently employed by a high-tech firm but was eager to "run something," in his words, and to get back into solar, his first love. I told him what I was up to and gave him a copy of my book, which had just been published, and at our next meeting he agreed to join my board and to help "scrub the numbers" in the business plan Gerry and I had been slaving over for several months. Standard Solar held its first board meeting, as required by Delaware corporation law, in the spring of 2005, in my basement office.

So now I had a legal "C" corporation, a board of highly qualified solar-industry veterans, and very little money. I'd raised some early "angel capital" from a woman in Boulder, Colorado, whom I'd met at a "sustainable resources" conference at the University of Colorado. Ellin Todd had told me, "If you ever start another company, I'd like to invest." She wasn't wealthy but had made a nice nest egg when a biotech company she worked for was sold, and she wanted to do good things with it. Ellin bought Standard Solar's first common stock, taking an enormous risk on an unproven venture started by an entrepreneur she barely knew. Her reward three years later was to get her money back fourfold. "You sent my daughter to college," she later told me. Ellin took me around to meet a number of Boulder's "socially creative millionaires," who talked a green streak about solar and saving the planet, but none of them would invest. "Why would we want to invest in a solar company in Maryland?" they asked me. These ecocentric trust-funders were also egocentric, I discovered, and not the type of people I wanted as shareholders. They lived in their own alternate, holier-than-thou, ecopure universe, which is what Boulder had become since I attended CU in the early 1960s, before it had become what Boulder's mayor called "twenty-two square miles surrounded by reality."

Boulder already had one new solar start-up, Namaste Solar, started by Blake Jones, who decided on the exotic name when he returned from volunteering as a rural solar installer in Nepal, shortly after we completed our SELF projects there. I went to see their new shop in downtown Boulder. "I knew about you in Nepal, and about SELF, and Pulimarang," he told me, "and I always wanted to meet you." I realized I was not alone in envisioning that the next "pure play" in solar PV in America would be professionally organized and managed solar-installation companies.

Blake had assembled a very young team (half the age of my group), and they were going for it. The demand for solar in Boulder was high, and they were there to meet it. They had organized their company as an employee-owned enterprise, without investors, just their own cash. Very idealistic. "We're all employee owners and we love it," he said. This was a far different model from Standard Solar's corporate/investor model. They had chosen a very strange name, I thought, "Namaste."

"Who the hell knows what that means if they haven't traveled in India or Nepal?" I wondered to myself. I wished them well, told them about our plans for Standard Solar in Maryland, and figured they wouldn't last; hell, maybe we'd buy them out one day.

The next time I saw Blake Jones was on page 2 of *The New York Times* standing next to President Obama overlooking a multikilowatt PV array atop the Museum of Natural History in Denver, installed by Namaste. By 2012, Namaste was the largest and most successful solar company in Colorado, with over one hundred employees and a second office on Larimer Street in Denver. "Namaste!"

Boulder itself would break new ground in 2011 when it formed its own municipal power company, got divorced from one of the nation's largest utilities, Xcel, and voted to "go solar" in a big way. The city then launched a $100 million "smart grid" project.

Back in Maryland, now with a little walking-around money, we finished our business plan and set out to raise more money. But before we did that, we decided we all needed to learn hands-on how to install an American solar home system. So, Lee, Gerry, a freelance electrician, an art school graduate named Matt Griffiths, and I practiced on a couple of Lee's customers. After fifteen years of running solar projects around the world, I had never actually installed a solar

Standard Solar technicians installing solar array atop Maryland townhouse. *(Author)*

panel, and I wanted to know everything about it: the wiring and cabling, the connectors, the mounting hardware, how to drill holes in asphalt roofs and seal them against leaks, and, most of all, how to stay safe and not get electrocuted (you can from PV) or fall off a roof. As a veteran mountain climber, I had no fear of roofs, the steeper the better, but I wasn't as agile as I used to be. I made everyone get harnesses and ropes to anchor possible falls. I made everyone wear hard hats before OSHA could fine us.

On our first jobs with Lee, we looked like the five stooges, old stooges at that, except for young Matt, whose job was to collect the cardboard from the solar-panel boxes for recycling, a task he managed with all the enthusiasm of a slug. I misjudged Matt, it turned out, since he eventually decided to forget about film school and an art career and became a dedicated, hardworking, technically trained and industry certified PV installer and one of Standard Solar's best and most dedicated crew chiefs. I could not have been more proud of him. Meanwhile, Gerry Braun went to California to meet two dozen solar pros and document—and work on—a residential system installation near San Francisco; we wanted to know how California's veteran installers did it. Lee and Gerry toured solar installations in New Jersey, which was fast becoming the second solar state after California, thanks to its new incentive programs.

In early 2006, Lee agreed to merge his LLC with Standard Solar, in exchange for stock. There was still enough money in the bank to sweeten the deal and convince Lee to quit his day job and work for Standard Solar. We soon found ourselves launching operations.

Rhone Resch, the tall, elegant-mannered, soft-spoken lobbyist who headed up the Solar Energy Industries Association, ran into me on Capitol Hill after a congressional "solar expo" and said he'd heard I was starting a company and told me he wanted to be our first customer. I was surprised he hadn't contacted one of the other long-established local solar installers. I never knew why he chose us, but serendipity had struck again; our new company could never have had a better advocate than the head of America's largest solarelectric trade organization.

He told us he wanted to use Sunpower modules, the new 22 percent efficient, all-black, 215-Wp modules that America's latest entry into PV manufacturing had just introduced. Because these would be located on a highly visible house in northwest Washington, which would be visited by federal and local government officials, as well as solar enthusiasts and the media, Sunpower gave us a good price. Rhone accepted our quote and applied for a grant from the D.C. government, which had just instituted a solar incentive program. Looking less stoogelike this time, the team directed by Lee all pitched in. Our electrician invented an electric lift to hoist the modules up to the very steep roof. I straddled the high-pitched roof like a bull rider, admiring the nice views of the leafy neighborhood while I helped drill holes and took photos of the momentous event.

I liked getting my hands dirty and learning the trade, but this is not what I'd set out to do. Company presidents aren't expected to climb on roofs and do manual work. In any case, I was supposed to be raising serious capital so this venture could go forward, so we could buy a warehouse full of solar modules at negotiated, factory-direct wholesale prices, enabling us to make solar affordable for customers and fend off any and all competition.

I had learned from SELCO's vice-chairman, John Kuhns, how the investment game is played, how to structure a stock offering, and how to close a deal. I learned there are only four sources of investment capital: friends and family, angels, venture funds, and private equity. Each has to be approached in a different way, and each has very different expectations and demands. By now I'd found five more angels, and our total capitalization was close to one million dollars. I retained control, but I'd diluted my own stock holding considerably.

We rented temporary offices, hired people, and, amazingly, the business poured in, mostly via Internet advertising since we couldn't afford anything else. Thanks to Google's "Ad Words," we found our market, or rather the market found us. I bought the book *Search Engine Optimization for Dummies,* from which I learned just enough to hire a professional expert in Internet and Google marketing. We framed a copy of our first $100,000 check, a deposit for a massive residential system on a hedge fund manager's home in northwest D.C. We expanded to new headquarters in an industrial park in Gaithersburg, ten minutes from my house. This was an easier commute than India had been. We hired more people, and more business came in, including solar hot water and some commercial PV systems.

Solar hot-water systems were a dilemma. The Chinese had transformed the solar water-heating business with the production of evacuated tubes, invented in Italy, using aerospace technology that allowed the sun to heat water quicker and hotter than the old flat-plate solar "collectors." The dilemma is this: unfortunately, most people don't know the difference between solar electricity and solar hot water, since both appear to come from "solar panels."

It was all "solar energy." I would explain facetiously, "You can tell the difference because one comes out of a faucet and the other comes out of a socket." In addition, one is an electrical business and one is a plumbing business. We tried both for a while, and did very well in the domestic solar hot-water business since it has a quicker payback and often saves more money for the user than solar PV, but it only deals with a small part of a home's energy needs. We subcontracted out the solar water-heater installations, and later got out of the business entirely. We were in the solar *power* business, not the solar hot-water business; not many companies do both well.

Parenthetically, an example of the misunderstanding of "solar energy" technologies is the issue of the famous White House solar panels installed by President Carter and removed by President Reagan. These were hot-water flat-plate collectors. Solarelectric panels were not restored to the White House proper until 2013, but George W. Bush had a 9.4-kilowatt PV array installed on an auxiliary White House maintenance building in 2002.

Our business plan focused on how we planned to brand, market, sell, install, and service residential solar home systems. We determined that to do this right and not just become another struggling home-grown enterprise, we needed substantial capital investment. This would allow us to prime the market, advertise, open a chain of solar service centers, expand to wider territories beyond Maryland, hire top people at good salaries, buy a fleet of trucks, and purchase large quantities of product up-front directly from manufacturers instead of through distributors. With enough capital, we could avoid the middlemen, which most installers relied on for their supply of modules and inverters.

Our staff continued selling and installing, while I started "dialing for dollars," making cold calls to everyone in my Rolodex that might wish to invest in our new solar company. We were seeking to raise $10 million while we simultaneously operated on our seed capital of $1 million. We forecast profitability in three years through rapid growth based on an assumption that our average sale would be for a 2.5-kilowatt system, with solar modules costing us $3.50 a watt. We were wrong! Soon enough, we were selling 4-, 5-, and 6-kilowatt residential systems, and commercial customers were coming in asking for 10- and 20-kilowatt systems, which seemed huge to us at the time. Nor did we know that, three years hence, we would be paying

$2.00 per watt for solar panels, and by 2012 they'd cost us less than $1.00 per watt! (And even less today.)

Despite our conservative assumptions, our hockey-stick growth projections, and our exploding business, investors weren't interested. I could publish a small book with the correspondence and e-mails from at least a hundred "green tech" venture capitalists telling us in great detail why this wouldn't work: silicon prices were too high (yes, but we knew they'd come down, and they did), people couldn't afford PV (they could, and we found ways to finance them), we couldn't get panels directly from producers (we did, right from the start), there were no government incentives (in Maryland, even without incentives, business was taking off), and, basically, solar was just too expensive. How venture capitalists stay in business with such clueless, unimaginative, and risk-averse managers is beyond me, since they were all wrong, every one of them.

We learned that venture funds, even the ones that touted their interest in the "clean energy space," weren't the right target, since they normally invested in new technology, not retail businesses, which we were. We pitched our plan to Ardour Capital in New York, which specialized in renewable energy, and they said no. We visited Kleiner Perkins on Sand Hill Road in Palo Alto, and they said no. I attended ACORE's Renewable Energy Finance Forum at the Waldorf-Astoria in New York and didn't meet a single investor interested in our business plan. I followed every lead I could unearth in the United States and Europe, including Frank Asbeck, CEO of SolarWorld in Germany, whom I had pitched in person, and got nowhere (he was right not to invest since solar manufacturers shouldn't also be in the downstream solar business). Frank, however, did believe as I did that America, in his words, was "the sleeping giant of solar" with the chance to reach grid parity before Germany because we had three times the sunlight. Frank would later build a 500-megawatt-capacity solar PV manufacturing plant outside Portland, Oregon, to supply the sleeping giant.

I flew to Vancouver to meet the billionaire mining magnate Frank

Giustra, a friend of Bill Clinton's who had told him he must invest in solar, and he said he'd put in $5 million if an American "lead" investor would match it. None would. Tony, our full-time CFO by now, pitched investor groups in D.C., Baltimore, and Philadelphia. Nothing. Gerry Braun told me he was getting "fund-raising burnout." He left, and took a job managing solar energy research for the California Energy Commission.

Some venture capitalists would ask, "Why don't you set up this business in California?" where solar was popular and financial incentives were available. I explained patiently that I didn't live in California, that Maryland and the National Capital Area had higher household incomes than anywhere in California, and that California already had many solar installers that had grown from mom-and-pops to successful small businesses. Venture capitalists never say no; they just rarely say yes, and they waste your time milking you for information. Many of those who turned us down later invested in similar solar-retail ventures, such as Solar City, which was launched the same year as Standard Solar with a $10 million infusion by PayPal founder Elon Musk (who also launched the Tesla electric car and founded SpaceX to supply the space station with his private rockets).

The Silicon Valley investment community and the state of California were throwing money at superstar "visionary" Bill Gross, the "dot-com king" who had started IdeaLab, which had spun off a new solar division, Energy Innovations. This company was started the same year we launched Standard Solar. Gross's team of thirty-five "solar geeks" in Pasadena had invented something they called the Sunflower, a PV-concentrator technology that Gross claimed could save the world "one rooftop at a time." He was featured in *Wired* magazine and many others, touting him as the man who would cut the cost of PV in half with his Sunflowers. These devices focused light on highly efficient PV cells through Fresnel lenses attached to aluminum boxes that were linked with motorized tracker devices. The first time I saw a photo of this contraption, I said to myself, "You've got to be kidding!" This may "work," I thought, and excite the eager clean energy investors on Sand Hill Road looking for "new,

Biannual Solar Decathlon featuring twenty university-built innovative solar homes, sponsored by U.S. DOE, Washington, D.C. *(Author)*

disruptive technologies," but it will never be practical, and certainly no homeowner would ever want such a thing on his roof.

In March 2012, the assets of Energy Innovations' photovoltaic assembly facility were auctioned off. Like Solyndra, the company was gone. Other "new technology" companies, like Amonix and Soitec, were still in business, planning or building multimegawatt plants in extremely sunny areas like Colorado, Arizona, Arabia, Mexico, and Australia, where concentrated photovoltaics (CPVs) may one day be practical. But read on.

Walking around the Solar Decathlon on the Washington Mall with Ed Begley Jr. one October afternoon, we were attracted by the model solar home with rows of black cylinders on its roof instead of PV panels. "That's PV," said Ed. "It's round, tubular, supposed to capture light from all directions, including rays bouncing off a white roof." Does it work? I asked myself, ever the doubter. I looked into it, called friends, and discovered that they were made by the darling of Silicon Valley's venture funds, Solyndra. I'd heard of the new firm, but not much. We all know what happened next, how the Obama administration made the unfortunate decision to approve the $543 million loan guarantee that had been prepared, but not finalized, under the Bush administration. It was on Secretary of Energy Steven Chu's desk when he arrived, and insiders informed me that he had been "ordered to sign off on that."

I later learned something else about Solyndra, besides believing like everyone else in the solar industry that their tubular PV "panels" were way too expensive to produce and technically impractical. One day, I was meeting with attorneys from the energy practice at one of the capital's largest lobbying law firms to talk about solar, and I volunteered my unsolicited opinion that funding Solyndra was a huge mistake. There was a brief silence until the chief energy counsel said, "We were paid ten million dollars over five years by Solyndra to get the loan guarantee through the DOE." The meeting ended abruptly, and my solar consulting services were no longer required.

Solyndra wasn't the last solar boondoggle.

In 2012 I visited the giant 30-megawatt Amonix CPV installation just north of Alamosa, Colorado, the largest of its kind in the world. Acres of massive seventy-five-foot-high towers support fifty-by-seventy-foot multiaxis tracking arrays of lens-focused high-performance photovoltaics. Developed and operated by energy giant Cogentrix, a subsidiary of Goldman Sachs, the huge contraptions became troublesome from the start, erupting in maintenance problems: too many complex moving parts and excessive water and power requirements to clean the Fresnel lenses. "This technology is doomed," I said to myself after looking in awe at this solar behemoth that filled the horizon. The plant was only three months old, and already it had been shut down much of the time because it didn't work. A week after my visit, I learned that Amonix, financed by several well-known venture investors, was forced to close its fourteen-month-old manufacturing facility in Nevada. An analyst at Greentech Media called it the "death rattle for the CPV solar industry." Shortly afterward, another CPV company, SolFocus, went under.

At the same time, a new CPV company, Semprius, opened a

Author with section of 210-kilowatt PV system atop the U.S. Dept. of Energy, Washington, D.C., installed by Standard Solar in 2009. *(Tony Clifford)*

CPV-manufacturing company in North Carolina. As Greentech Media commented, "Talk about going against the flow!"

Meanwhile, multilevel marketing organizations like Citizenre were raising millions by selling dreams of delivering cheap solar with panels from their own factory. James McKirdy, called a Florida-based "solar grifter" by industry analysts, bilked investors out of millions by offering his "breakthrough technology": NXGen "electronically produced" solar panels delivering 46 percent efficiency, double the conversion rate of the highest-rated commercial modules. Brooklyn Bridge, anyone? When I hear the word "breakthrough," I flinch.

I began to understand the thinking of many investors: better to lose money on something new and sexy than to make money on something old and boring. Makes for more interesting cocktail party conversation, I guess.

Well, at least we had our angels, and thanks to them we established what quickly became the biggest solar sales, service, and installation company in the D.C. area. Our growing sales team kept signing up residential and commercial customers faster than we could serve them. Lee Bristol got our local utility, PEPCO, to buy a 20-kilowatt system for the roof of a substation. On a cold December day, twenty of us bundled up to install the system in howling subzero winds atop the five-story building, loading and unloading crane-hoisted pallets, assembling the mounting racks, wiring the big inverters. Tony then convinced the Department of Energy to purchase for cash a 205-kilowatt system for the roof of its Forrestal Building on the Washington Mall, where I had worked for Tom nearly thirty years earlier. While they worked on the DOE project that summer, our technicians enjoyed sweeping views of America's most beautiful city. This was our biggest installation to date, which would later be eclipsed by much larger systems in the Washington area.

From 2006 through 2008, we were focused on residential solar, which had the biggest profit margins. First, we had to find out how to "sell solar" to homeowners, one at a time, who were phoning us at the rate of up to five hundred calls a week. We trained a phone bank. Lee was a master of explaining how much PV a homeowner could fit on available south-facing roof space (which we could see from proprietary aerial imaging, before Google Earth was around), how much power the system would produce, how much it would cost, and what the "payback" might be. I say "might be" because "payback" is determined by two things: the rise in electricity rates, which can't be predicted except to know they will continue to go up, and what financial "incentives" were available that would help reduce the up-front cost to the homeowner or business—incentives also can't be certain since they change yearly, from state to state and county to county. We explained that after a time—ranging from five to eight years—the system would have paid for itself and would then produce free electricity for the life of the panels. Most manufacturers now guaranteed their panels for twenty-five years (experts think they will last much longer). Meanwhile, the money you were paying for the system was buying something you would own, not going into the utility's pockets.

After 2008, we could compute in the price the 30 percent federal tax credit that Senator Maria Cantwell of Washington had attached to Bush's "bailout" bill with the support of Representative Gabrielle Giffords of Arizona, one of Congress's most ardent supporters of solar energy (later shot and nearly killed by a nutcase). "Ever since I was first elected to Congress in 2008, advancing solar power has been one of my highest policy priorities," she said in a video interview. This consumer tax credit was the largest ever granted by the U.S. Treasury, revolutionizing the American solar industry overnight. It expires in 2016.

In Maryland, after we repeatedly testified in Annapolis in favor of state solar tax incentives and grants, the state, followed by several counties, passed progressive measures in support of residential solar. Delaware enacted similar legislation. The District of Columbia already had a generous but problematic solar grant program in place. I won't describe all these in detail; up-to-date information can be found on the Web (see Appendix 2). But by 2010 homeowners were enjoying, in some markets like Maryland, Colorado, and New Jersey, a 50-percent-off sale in photovoltaic home-power systems, and hundreds of thousands of families nationwide took advantage of the opportunity to "go solar."

I wished our company could have expanded fast enough to grab more of this exploding market, but we still had no new expansion capital to draw on, and, besides, we had our hands full meeting the demands for solar in Maryland, D.C., and even Virginia, with its much lower electric rates and no state or county incentives. Many people wanted solar, regardless of cost, and they beat a path to our door. I had always believed this would be the case: build it and they will come, and they came. We proved supply creates its own demand.

I had also believed that electric rates would continue to go up and that PV costs would continue to go down, and at the crossover point there were billions to be made. Somehow, I could never get this point across to the young financial analysts at the venture capital and institutional investment firms we pitched. But we no longer cared what fund managers thought. We were busy "making solar standard," our new motto, emblazoned in blue on the back of our technicians' bright yellow T-shirts.

* * *

Why were all these homeowners and businesses buying solar? I'm a lousy salesman myself, and rarely closed a sale when people happened to get me on the phone. But to understand sales and marketing, I needed to know what the motivators were. So I traveled around our region and visited our customers (and took photos of their installations, which the installers themselves never had time to do) and usually asked the husband, and sometimes the wife or a single householder, why they opted to make the investment. Back then, a solar system could cost, without grants or tax credits, $35,000 to $50,000—more if they went whole hog and chose to supply all their power from solar. Most were happy to get half their total kilowatt hours from the sun. For example, a home with 350 square feet of south-facing roof could hold twenty-four, 175-Wp panels (now much larger panels are the norm), producing, at D.C.'s latitude, 5,157 kilowatt hours a year, or 430 a month. The average American household uses 1000 to 1200 kilowatt hours a month. And once people plugged into the sun, they began to *unplug* everything else they could think of and consciously began to conserve energy and install energy-efficient lightbulbs and appliances. Even with minimal financial incentives, the overall return on investment in an SHS was between 13 and 15 percent; you can't get that at a bank, and rarely in the stock market, and the principal can't go down!

"Our family decided to go solar to reduce our greenhouse-gas emissions and make our home more eco-friendly, " said Mathew and Patience Bryan-Taff of Silver Spring, Maryland. Their modest ranch-style home belied the notion that only the rich could afford solar. Another suburbanite told me, "The panels are the envy of my neighborhood as we now get power bills twenty-five percent lower than last year. We look at our wireless monitor to see how much electricity we're producing and how much CO_2 we have not produced." Bob Thomason was talking about his 2.5-kilowatt system in North Carolina, where we later installed a 450-kilowatt rooftop array on the Terry Sanford Federal Building in Raleigh.

We calculated that the average 4.2-kilowatt PV system would reduce a household's carbon footprint by 164 tons over the thirty-year life of the installation. Solar was going to play a much bigger role in

America in helping to reduce greenhouse-gas emissions than it ever could in the developing world because Americans, on average, used at least one hundred times more electric power per household than a farmer in India. The Indian family was happy with a 50-Wp panel supplying all of their lighting, plus power for TV and radios, while a suburban family in Maryland had to purchase one hundred times more solar watts to cover even its minimum electricity requirements. Clearly, residential solar in the United States was going to be a much bigger business than selling solar in the two-thirds world: we had sold solar home systems in Sri Lanka for $400 but in America were now selling SHS for $40,000! And we were reducing carbon emissions by a much bigger factor than by merely retiring a couple of kerosene lamps.

Rhone and Lisa Resch, our first customers, told *The Washington Times,* "The primary reason we 'went solar' was to do our part to protect the environment. In D.C. the greatest impact on air quality comes from the use of fossil fuels to generate electricity." Rhone added, "By going solar we were able to reduce our carbon footprint while also reducing our monthly energy bill. We also wanted to make

Maryland colonial home with 7-kilowatt PV system produces more daytime power than the family uses; excess electricity is fed back to the grid. *(Author)*

Standard Solar's first customer: 7-kilowatt PV system covers nearly all the power needs of the Resch family in Washington, D.C. (*Author*)

choices that positively affect our environment and improve the country's energy security." Rhone told me later, "The news story didn't tell the half of it. You have got to come over on a sunny day and watch our electric meter flying backwards. It is amazing! Yesterday, we started producing power at 7:35 A.M."

One day Rhone and I went to lobby our congressman, Representative Chris Van Hollen, and Rhone pulled out his BlackBerry (this was before iPhones) and brought up a chart that showed his home's real-time electricity consumption. It was midday, and his 7-kilowatt system was sending power to the grid. Then, suddenly, the graph reversed itself, and his home started using more power than his panels were producing. "Oh, that means my wife just started the washing machine or dryer," he told Van Hollen, who stared, fascinated, into the little screen.

Our "early adopter" customers often did a great deal of research on their own before calling the company. Dan Redmond of Arlington, Virginia, had visited dozens of houses on the Washington area's annual Tour of Solar Homes, sponsored by the American Solar Energy Society (ASES). He attended the local Green Festival and stopped by our solar booth, and he had also visited the second Solar Decathlon on the Washington Mall in October 2005, sponsored by the U.S. Department of Energy and BP Solar. This biennial event creates a veritable suburban tract of twenty all-solarelectric houses, which are built by engineering and architectural students at their respective invited American and European universities and shipped to Washington to be reassembled in front of the Capitol. Over one hundred thousand visitors show up, and it is a sight to behold. That year, the house sponsored by my alma mater, CU, won first prize for its overall energy performance.

Dan knew what he wanted, and why. He and his wife had grown up in the coal country of West Virginia and knew the damage coal mining and mountaintop removal was doing to families and communities, not to mention the pollution from coal-fired power plants that wafted over Washington, which has the highest concentration of greenhouse gases of any city in America (cars are partly to blame). The Redmonds also wanted their children to learn about technologies that offered hope for the future, and bought them solar toys to demonstrate how PV worked. Dan wanted to be a beacon for his neighborhood, and since his available sun-facing roof was visible from the street, he asked us for the new, aesthetically pleasing, all-black Sunpower modules, which were also the highest-efficiency production panels on the market (22 percent) at the time (since eclipsed by Trina Solar and Panasonic-Sanyo). Their high sunlight-to-electricity conversion rate is thanks to the lifelong laboratory work of the visionary scientist Richard Swanson, who founded the company and sponsored one of the Decathlon houses. We had just become a "premier" Sunpower dealer, after Gerry and I flew to California to convince the company, which had just gone public, raising $140 million, to sell to us directly.

Dan got his modules and later told me how much he loved living "quietly." "Besides producing pollution-free electricity on-site, the system is virtually maintenance-free. During pollen season, I use a

Suburban Virginia couple with their Standard Solar PV and solar hot water installations. *(Author)*

garden hose to rinse off the modules. Otherwise, the system's operation is effortless. It starts generating electricity automatically as soon as sunlight hits the PV array and goes off-line at dusk. It's that easy. And that quiet."

We offered customers several choices of panels, including BP Solar's cobalt-blue polycrystalline panels, manufactured thirty-five miles up the road from us, in Frederick, Maryland. Gerry still had connections there, and we convinced the president to break all precedent and sell to us directly, instead of through their established nationwide distribution channels (middlemen). These panels were not as large or efficient as Sunpower's, but they could be installed more quickly through an ingenious mounting system that allowed them to be bolted directly to the roof in a low-profile array. And some people liked the shimmering blue flash of the polycrystalline cells.

We incorporated "just-in-time delivery" from BP's warehouse, provided we preordered hundreds of panels, something the mom-and-pop solar installers could not do. This was in 2006–07, ages ago in the rapidly changing landscape of global solar photovoltaics production.

The real solar sales driver in those days was global warming, something no one doubted at the time. This was well before Fox News, the coal and oil malefactors, and the Republican Party had brainwashed millions into believing global warming and climate change were a "hoax" or, worse, a conspiracy. No one at the time could have imagined the Republican Party in 2012 would field a half dozen presidential candidates that refused to take climate change seriously and would ridicule Al Gore every chance they got.

In 2006 Gore's movie *An Inconvenient Truth* was released, and shortly afterward, his book. Gore won the Nobel Prize and an Academy Award for his efforts to educate the masses once and for all (we thought) about mankind's contribution to the warming of the earth and the changing of its weather. More books, like Tim Flannery's *The Weather Makers* and Mike Tidwell's *The Ravaging Tide,* found wide audiences. No one could read these books and not believe we have a problem, which is that the continued and growing use of carbon fuels will one day kill us. Flannery, who advocates completely "decarbonizing" our electric grid, wrote, "I became so outraged at the irresponsibility of the coal burners that I decided to generate my own electricity, which has proved to be one of the most satisfying things I've ever done. For the average householder, solar panels are the best way to do this."

Former *Washington Post* journalist Mike Tidwell had put solar panels on his Takoma Park home in the early 1990s and held regular open houses to show it off. He launched the Chesapeake Climate Action Network, one of the most active community organizations in the country fighting global warming. He also promoted Standard Solar at his monthly solar-home open houses. His first book, *Bayou Farewell,* predicted the Katrina disaster years before it happened, based on studies of extreme climate threats from the rapid warming of the earth—it is warming faster than it ever had before man.

Former assistant energy secretary Joseph Romm published *Hell and High Water* in 2007, in which he notes that we don't need "new" technology, just the *political will* to use existing clean technology, which, he says, "over the next two decades . . . could cut U.S. carbon dioxide emissions by two-thirds without increasing the total electric bill of either consumers or businesses." Romm vigorously criticized the policy makers' favorite delaying tactic, which he called the "technology trap": always waiting for "technical breakthroughs."

At least our customers weren't waiting for "breakthroughs" or lower-cost solar cells; solar was here, now, and they wanted it, largely to reduce their carbon footprint, but also to hedge against rising utility rates. In addition, some customers opted to purchase battery storage systems that would keep their grid-tied solar working in the event of a power outage. The bank of industrial lead-acid batteries (stored in a vented basement or outdoors in a shed) is connected to an "emergency" circuit that will keep lights on, a refrigerator working, and power one outlet for a TV, a computer, and to charge cell phones. They could survive quite comfortably with their off-grid sun-power system.

Ironically, the "solar solution" that helped reduce global warming would also protect families from the results of global warming: climate catastrophes that plunged millions of utility customers in the dark.

In those years, every magazine at one time or another featured global warming on its cover. One *Time* cover proclaimed, "Be Worried, Be *Very* Worried," while another promoted its "global warming survival guide." *The Economist, Outside, Wired,* and *Vanity Fair* all published their "green" issues, and a *Newsweek* cover displayed a photo of California's governor Arnold Schwarzenegger holding a globe with the headline "Save the Planet *or Else*." This was well before a collective backlash by vested interests and elected imbeciles pushed environmental concerns right off the table. Denial is easier than doing something.

I put all these magazine covers together in a collage poster that asked in bold black letters, "Isn't it time you did something? You can!" Below was our logo, the words "saving the planet one roof at a time," and Standard Solar's phone number. I was distressed I couldn't put solar on our own house, but our steep, shake-shingled, multigabled roof rendered it all but impossible, even though our "new urbanist" community, at my urging, had passed a resolution in support of solar home systems. At least we bought a Prius, as Tony, Lee, and Gerry had already done; it was something.

Not everyone could afford to "save the planet" by such actions, for even with multiple incentives, grants, and the easy finance programs we offered, solar was still costly for most families. We later introduced solar leasing, as our competitor Solar City had done. With a solar lease, customers could lock in the cost of their electricity for fifteen or twenty years without any up-front cost—this became a billion-dollar business in California as third-party financiers like Google and Bank of America entered the solar home and commercial markets by partnering with companies like Sunrun, One Roof Energy, and Sungevity.

Sunrun finances residential customers for companies like Standard Solar. One Roof Energy raised $100 million in 2012 from U.S. Bancorp and the Korean industrial giant Hanwha to finance four thousand home-solar customers. In 2013, SolarCity raised a half-billion dollars from Goldman Sachs to finance solar roofs.

Sungevity's founder, Danny Kennedy, another Greenpeace veteran, led the effort in San Francisco to float a successful ballot initiative that resulted in the city issuing $100 million in revenue bonds for the public finance of solar power systems. Then, in 2008, he launched his California residential solar company, which today has 260 employees. He also wrote *Rooftop Revolution*, published in 2012, an indispensable guide to rooftop solar.

While our business grew 100 percent a year, and while we would eventually reach our thousandth residential customer, this was only a grain of sand in the great beach of the potential American solar market. No one company could ever serve it all, but the good news was that many were trying. Today, it is estimated that there are over two thousand solar energy sales-and-installation companies in America. Many were started by young people with little or no experience in solar electricity; they just had a burning desire, so common among their generation, to spend their lives making a living by doing something useful. Solar was a way. Soon, we had local competitors, but they were welcome since we all knew one company could only scratch the surface of the growing sun-power market.

By 2008 we had eighteen employees and several part-time licensed electricians. I had handed over the CEO responsibilities to Tony Clifford, and we hired a CFO. A dedicated cadre of enthusiastic young people had joined the team, many with no knowledge of electricity, but quick to learn. Many had advanced degrees and were educationally overqualified for the blue-collar positions we had available. Most believed in living the ethical life and doing good. Marketing, selling, and installing solar was always more than just a job. It was enduringly satisfying. And, we paid well, although everyone put social concerns above the desire to make money.

Occasionally I, myself, would bring in a customer. After an energy symposium at the National Press Club one day, Bob Freling introduced me to Thomas Friedman, the *New York Times* foreign affairs columnist. He had written extensively on renewable energy and "global weirding," including a recent *New York Times Magazine* cover story on "The Power of Green," reporting on how America could rebuild via a green economy. "Embedding clean-tech into everything we design can revive America as a manufacturing power," he had written. Friedman had recently interviewed Bob about SELF for his upcoming book, *Hot, Flat and Crowded*. Friedman had proposed in one of his columns that the president call on "every U.S. school to raise money to buy solar-powered lightbulbs for every village in Africa that didn't have electricity so African kids could read at night."

"Tom, I'd like you to meet the founder of SELF," Bob said. Tom didn't seem very interested. "Now he's running a company in Maryland to bring solar to American homeowners," he added. Tom brightened. I told him, "I figured if we could do it in India, we could do it here."

Tom said, "We've been looking for a solar company to put solar on our home in Bethesda. But it has a slate roof. It's been featured in *Architectural Digest,* so we don't want anything very visible. We don't know where we can put solar panels." I said we could help. "I'll e-mail you this afternoon," he said. By the time I got back to my office, his e-mail was there with his telephone number. I called him and set a time to meet at his thirteen-acre Bethesda, Maryland, estate. It was among the finest houses I'd ever seen in the D.C. area.

Tom was very down-to-earth about the problem, which was that they used too much electricity in the huge home and he needed to do something green so he could practice what he preached. But where? Lee and I toured the property. The slate roof was out, and a ground array visible from the road or driveway was vetoed by Tom's wife. That left the side of his property, next to but obscured by the house. There, we installed 7-kilowatts of PV on two pole mounts after petitioning the Montgomery County Council for permission to erect a structure in a side yard. That zoning change took six months, but in the end Tom was pleased that he was "doing his part." We told the couple they really needed to do an energy audit and begin to conserve on their energy consumption if the solar panels were going to make any difference.

New York Times columnist Tom Friedman's 7-kilowatt pole-mounted solar power system alongside his home in Bethesda, Maryland. *(Author)*

I remained the company's chairman, still working out of my basement office at home. I wasn't needed at headquarters; I would just get in the way. Everyone, from managers to accountants to telephone salespeople to technicians knew exactly what they were doing. In any case, I never had any intention of running a company, just starting one. Now, it was started and operating successfully, but I still hadn't raised the capital we'd set out to raise: $10 million. No investors were interested beyond our patient angels. We weren't strapped for operating capital and could have kept right on "bootstrapping" the business, as it were. But we could not have realized our ambitious growth projections, or achieved the social and environmental impact we envisioned, without additional funds.

During several years of attending the Clinton Global Initiative in New York, I met numerous billionaires touting their interest in "renewable energy technology," but there were no takers when it came to a retail business that would actually sell and install these technologies. They were all looking for the Silicon Valley megareturns that they had experienced in the dot-com era (and we all know how that greed binge ended). Solar didn't need that kind of investor anyway.

Then, one day, out of the blue, I got a call from a private equity firm in Middleburg, Virginia, run by two fifty-year-old corporate veterans. The lead partner, Steve Lamb, was a West Point graduate with a Harvard MBA and extensive credentials running Fortune 500 companies (Dresser Industries and Case International, the latter which he took public). He had made his money, bought his Virginia manor house and horse farm and his Wyoming mountain retreat, and now wanted to be an investor, putting some of his and his friends' capital to work. One of our employees had given him our business plan.

"This is the best business plan I've ever read," he told me on the phone. "When can we meet?" We set a date, and he, along with his partner, Jim Sharman, also a West Point grad with an MBA from Duke, showed up at our Gaithersburg offices, and we sat around our used conference table in our secondhand office chairs. "We'll put in two and a half million now, and another eight to ten million later." We didn't believe them. "Stop shopping the plan," Steve said, after I told him we had a couple of potential players looking at the deal. "We can't just stop our fund-raising efforts like that," I replied. We asked for a nonrefundable hundred grand for an exclusive option, and they agreed and quickly transferred the funds while everyone got down to their respective due diligence responsibilities.

These were private equity investors who regarded the venture guys with the same disdain we did. They knew nothing about the solar industry, but they knew a lot about business and how to grow small corporations into very large concerns. Although they were Democrats and Obama supporters, they weren't "green" investors; they saw solar as the next big thing and wanted to get in early and make a bundle. Standard Solar was a golden opportunity. Private equity always takes control, never a minority position, and that means they put in their own people from the get-go. I was bought out at a very fair price, keeping half my common stock. Steve took over as chairman and kept Tony as CEO.

They paid the share price we'd set, accepted our "valuation," and

Mark Ohrstrom, Standard Solar investor, atop his solar-powered office building in Middleburg, Virginia. *(Author)*

acquired 67 percent of the company by buying up most of the angel-investor shares. Everyone was happy. I was pleased there would only be one class of stock in the company so we were all in the same solar boat, so to speak. True to their word a few months later, in 2008, they raised and invested another $8.5 million—the same week Lehman Brothers collapsed. After Solar City, Standard Solar was now the best-capitalized solar-installation company in America.

I was now seven years older than my father had been when he retired to the Carolina mountains from the manufacturing company in Cleveland where he'd worked for thirty years. I was happy to relinquish my responsibilities while I stayed on the board. They brought in an ex-army officer and systems engineer to manage day-to-day operations and report to them. He had no solar experience but learned quickly. Steve and Scott, the new president, put solar on their respective homes, and Tony would follow. The company would be run very differently now, with apparent slackers fired without warning, and no more touchy-feely team spirit. But they got results, hired more and more people, fine-tuned sales and marketing, bought more trucks, and contracted for megawatts of PV inventory. Standard Solar's shared values were no longer based on saving the planet. The new mission was making money. And they did.

While the residential business was initially the cash cow, large high-dollar projects started to come in: a 700-kilowatt array atop the county's largest ice rink; hundreds of kilowatts at the National Archives in Silver Spring, Maryland; 2,600 PV modules installed at Catholic University; a large PV system at American University in D.C.; 850 kilowatts on the University of Delaware's field house; a 1.5-megawatt system at the Perdue chicken farms on Maryland's Eastern Shore; and a 120-kilowatt parking structure at Washington and Lee University that eliminates 114 metric tons of carbon dioxide every year. Three utilities on Maryland's Eastern Shore purchased multimegawatt arrays financed by Washington Gas Energy Services. Standard Solar covered a parking lot at Anne Arundel Community College with a 763-kilowatt solar system, the largest carport structure on the East Coast. Baltimore's utility bought a large system to cover a brownfield next to the Ravens' Stadium. Federal contracts

Standard Solar's 850-kilowatt solar installation atop the University of Delaware's field house. *(Standard Solar)*

rolled in. These, and many more, even larger, commercial and institutional solar installations would follow over the years as the company divided itself into commercial and residential components. The company had come a long way since we installed the twenty-Wp system atop a PEPCO substation that seemed so large to us at the time. By the time you read this, most of these systems will have been dwarfed by many more in the pipeline.

In 2010 *Inc.* magazine named Standard Solar the seventy-third-fastest-growing company in the United States with over 80 percent year-on-year growth (it remained on the list in 2011 and 2012). The company had moved to a new seventeen-thousand-square-foot headquarters in Rockville and had over one hundred employees. For fun, I started photographing our customers' installations, using Google Maps to track down hundreds of our customers' homes scattered all over the region, from modest one-story ranch homes to huge colonial mansions with their street-facing roofs covered entirely with solar panels. Six- and seven-kilowatt systems were common now.

The company's revenues hit $40 million-plus and the business was profitable. Utilities were hiring Standard Solar to install multimegawatt projects, and more and more businesses were opting for large solar-rooftop systems. The biggest boost for solar was the unexpected 50 percent price drop in 2011 for PV modules, mostly because of cutthroat competition from China (see chapter 12). Standard Solar began buying from Chinese manufacturers such as Suntech and Trina Solar. The company's proudest accomplishment at this writing, beyond installing 23-megawatt and generating 42,000-megawatt hours of clean power, is reducing CO_2 emissions by over thirty thousand metric tons, according to the EPA's "green-power-equivalency" calculator.

I left Standard Solar's board in 2010 and moved to Florida, knowing that the company was in good hands and that it would prosper, despite growing competition in the retail solarelectric business—even Home Depot had begun selling solar! It would continue to be entrepreneurs who would play the most effective and important role in the American solar industry, the more the merrier. I was glad to have been one of them and hoped to inspire—and be inspired by—others who were getting into the game. My advice to anyone who may wish to pursue a similar career path is this: if an inexperienced, middle-aged former journalist without specialized academic credentials, a background in business, or any money can launch an enterprise like SELCO and later a growing company like Standard Solar Inc., almost anyone can. *You just have to want to do it.*

In the most difficult times, I was often inspired by the words of W. H. Murray, a Scottish Himalayan climber, which I had framed on my wall:

Until one is committed there is hesitancy, the chance to draw back, always ineffectiveness. Concerning all acts of initiative (and creation), there is one elementary truth, the ignorance of which kills countless ideas and splendid plans: that the moment one definitely commits oneself, then Providence moves too. All sorts of things occur to help one that would never have otherwise occurred. A whole stream of events issues from the decision, raising in one's favour all manner of unforeseen incidents and meetings and material assistance, which no man could have dreamt would have come his way.

I also took heart from Goethe's couplet, which had guided me for years:

> Whatever you can do, or dream you can, begin it.
> Boldness has genius, power and magic in it.

And then there is Einstein, who reportedly said:

Everything is energy and that's all there is to it. Match the frequency of the reality you want and you cannot help but get that reality. It can be no other way. This is not philosophy. This is physics.

Perseverance and persistence always win the day.

12

America is now the world's fifth-biggest solar market, after Germany, Spain, Italy, and China. More than a quarter-million U.S. homes get a substantial portion of their power from the sun. We have over 8.5 gigawatts (GWs) of installed photovoltaics on residences, commercial buildings, institutions, government offices, military installations, and utility-scale solar farms. This is the equivalent of eight average coal-fired power plants. We have a long way to go, but sun power will eventually replace nearly all our 492 coal plants one day, most of which are more than forty years old.

In 2012, the United States installed 3.3 gigawatts, not even close to Germany, which installed 3 gigawatts of PV in just one month the previous year, but not bad. In fact, PV sales in 2013 grew 89 percentage over 2012 in the U.S. It is difficult to keep up with the fast-moving world of renewable energy here and abroad. China installed 7 gigawatts in 2012. Meanwhile, its factories turned out 32 gigawatts of PV modules in the first quarter of 2012, mostly for export.

In the U.S., residential demand spurred the largest growth. Worldwide, two thirds of global PV has been installed in the past two and a half years.

[A little refresher here: 1,000 watts is a kilowatt (kW), 1,000 kilowatts is a megawatt (MW), and 1,000 megawatts is a gigawatt (GW). One kilowatt in the developing world will light up twenty-five homes. In the U.S., the average American residential installation is now 5 kilowatts. A 1-megawatt PV array will provide most of the power needed by two hundred homes when coupled with energy efficiency and conservation.]

The solar revolution has come about because of the confluence of several events: the explosion of Chinese PV manufacturing, reducing the retail price of solar power by more than 70 percent; the success of the German solar program, which spurred enormous growth of PV sales; rising electric rates following America's utility deregulation; President Obama's economic recovery program, which plowed over $65 billion into renewable energy, creating more jobs than in any other industry; and federal, state, and county solar incentives, driven by concern about greenhouse gases and what carbon fuels are doing to our planet. This is all good news that I couldn't draw on a decade ago to answer "why aren't we using solar?"

But it's not all good news, because we could be doing more, with even greater benefits for the economy, job growth, the environment, and our collective future.

I'm reminded of the woman out on a shopping trip who calls her husband and tells him, "I have some good news and some bad news. Do you want the good news first, or the bad?"

Her husband says, "Give me the good news first."

"Well," she says, "the air bags work."

The good news first is that solar and renewables *work*; they're here now, growing faster than anyone can document; and they will be earth's safety device. The bad news is that we've already had the car wreck. Great damage to the planet has already been done, and there is going to be hell to pay before we reach the inevitable outcome of the solar revolution, a 100 percent carbon-free, sustainable energy economy. This revolution will be like the French Revolution, which "worked," but not before two hundred thousand innocent French citizens paid the ultimate price for change. The same with war—peace eventually comes as it must, but not before millions die unnecessarily.

Back to the good news.

According to Bloomberg New Energy Finance and their research partner, the Pew Clean Energy Program, global investment in renewable energy hit a new record of $263 billion in 2011, a 600 percent increase since 2004. Solar photovoltaics comprised $136 billion of that, $45 billion of that in the United States, which doubled that number in 2012; the rest is mostly going to wind, solar thermal plants, biomass, biofuels, methane, waste-to-energy, and solar hot water.

These figures don't include the rapidly expanding sectors of energy efficiency and conservation. According to a LinkedIn Analytics study for the president's Council of Economic Advisers, the American renewables industry grew by 49.2 percent between 2007 and 2011. In contrast, the Internet and digital media industry grew by 24.6 percent. All these numbers will be much higher by the time you read this.

My friend Jigar Shah, founder of SunEdison in Maryland, which quickly became the largest installer of commercial PV in the world for several years, revealed an astounding fact: 75 percent of all incremental energy as of 2012 was from noncarbon sources. In other words, the largest source of energy growth in the United States now comes from renewable energy. Nearly half of all new electricity installed in 2013 was solar.

The fastest growing of all, as I've previously mentioned, is PV, which has grown 80 percent each year since 2010. If you haven't guessed already, I'm partial to PV—which this book is about—because it has no moving parts, requires no fuel, and emits no carbon other than during its initial manufacture. By 2010, of the 565 gigawatts of clean energy online worldwide, over 65 gigawatts of that was solar photovoltaics. The figure is much higher now, *more than double the global installed-nuclear-generating capacity.* PV will rule the energy world as it turns to renewables over fossil-fueled and nuclear power generation. According to Greentech Media reports, approximately twenty-seven nuclear power plants' worth of solar was installed globally in 2012. The United States installed 723 megawatts of new solar capacity in the first quarter of 2013.

But King CONG (Coal, Oil, Nuclear, and Gas), still rules, and will for a long time. First, electricity generated from coal is the world's largest business, followed by the oil industry. Big coal and the Koch Brothers' super-PAC spent millions attempting to defeat President Obama in 2012.

The fossil-energy behemoths have grown in just over one hundred years from Edison's Pearl Street generating plant in New York City and from Rockefeller's oil derricks in Pennsylvania and Ohio to become the largest and most powerful enterprises the world has ever known. Some electric utilities like Électricité de France, South

Africa's Eskom, Tokyo Electric, Ohio's AEP, International Edison, Holland's Nuon, and Germany's RWE are wealthier than many countries; few armies are as big and powerful as ExxonMobil or Royal Dutch Shell. In fact, after sex, the most powerful force on the planet is the global oil industry.

Against this panoply of power led by the fossil-fuel theocracy march the forces of sustainable, renewable, clean, carbon-free energy. Sometimes it looks like a war, but it need not be. While many vested interests push back hard, buying politicians and funding "climate deniers" and launching public disinformation campaigns, many large utilities and energy corporations are nonetheless adopting clean-tech solutions.

Florida Power & Light, Duke Energy, Southern California Edison, Austin Energy, Constellation, PG&E, LADWP, AEP, NRG, NYPA (see glossary!), and many other utilities are buying all the solar they can get (and wind power), either because they have to as mandated by state renewable portfolio standards or because they want to because they see it is the future of energy in America. Utility-owned solar increased 300 percent from 2009 to 2010. America's biggest power provider to U.S. utilities, NRG Energy, Inc., has begun installing residential and commercial rooftop solar directly. Duke Energy sees solar as a potential threat and is considering entering the rooftop solar business itself.

However, as of mid-2012 we were seeing a "pushback" from some utilities against solar net metering and independent power producers as they began to see solar eating into their once monopolistic electricity markets. *The New York Times* reported in 2013: "Alarmed by what they say has become an existential threat to their business, utility companies are moving to roll back government incentives aimed at promoting solar energy. . . . At stake, the companies say, is nothing less than the future of the American electricity industry."

Arizona's largest utility asked the state power commission to reduce the value of net-metered solar credits and to charge residential solar users a large monthly hookup fee. California's three major utilities estimate residential solar is costing them $1.4 billion a year in lost revenues. In some states, where the public utilities commissions are in fact a fourth branch of government, the regulated power monopolies are pushing to end net metering. It's a battle not unlike the battle faced by AT&T years ago when the Baby Bells and independent phone companies ended their monopoly. Decentralized power production and "DG" (distributed generation) will revolutionize the electric power markets. It's inevitable, but dinosaurs don't die easily. Stay tuned.

Nonetheless, the Southern Company, the huge Georgia utility known for its active opposition to solar, built a 30-megawatt photovoltaic array in New Mexico in 2011. Amazingly, it was the Georgia Tea Party that urged the state utility regulators to force Georgia Power to use more solar energy. In 2012, Georgia Power solicited bids for 210 megawatts of solar electricity. NRG Energy and First Solar completed their 250-megawatt Agua Caliente PV plant in mid-2012. Oil giant Chevron is building solar thermal and concentrated PV plants, and no one is making them do it. In 2012 GE partnered with NexEra to buy the 550-megawatt Desert Sunlight solar project in California.

Bechtel and Brightsource Energy, along with NRG and Google, just completed the world's largest solar thermal plant in the Mojave, the $2.2 billion, 377-megawatt Ivanpah Solar Electric Generating System in California. Bill Clinton visited the site in 2012 and was astonished to find 2,100 workers on the job installing 53,000 heliostats that focus the sun's rays on a giant steam-generating boiler atop a 350-foot "power tower."

Another 110-megawatt solar plant is going up in Tonopah, Nevada, financed in part by a $737 million loan guarantee from the DOE, secured by a power purchase agreement with the Nevada Power Company. The project boasts the largest "power tower" in the world, at 640 feet, plus 17,500 heliostats to heat molten salt. Even Indian tribes are getting into the game: Paiute Indians allocated two thousand acres of tribal land for a 350-megawatt PV project thirty miles north of Las Vegas, which they will own.

In California, NRG Solar and MidAmerican Solar, along with First Solar and Sunpower, are building two of the largest PV arrays in the world at the California Valley Solar Ranch and at Topaz Solar

Farms, ten miles apart on the Carrizo Plain one hundred miles north-west of Los Angeles; their combined 800 megawatts to be purchased by PG&E came on line in 2013.

Big corporations have joined the solar revolution. Dow Chemical has gotten into PV, promoting their "Powerhouse" solar roofing shingles, while 3M's solar division supplies critical materials for PV manufacturing and Firestone's roofing division is now offering solar solutions. Lennox, a manufacturer of air conditioners, is advertising Solar Home Energy Systems on national TV. Westinghouse Solar makes panels. European firms such as Bosch, Alstom, Schneider Electric, ABB, and AEG are aggressively entering U.S. solar markets as manufacturers and project developers. As mentioned, Bechtel, AES, and Chevron are building huge solar plants. Cleveland's Eaton Corporation is manufacturing big inverters. I've already high-lighted the major Japanese firms and China's.

As everyone knows, nearly every auto manufacturer worldwide is offering an electric or hybrid-electric car that could one day result in the greatest energy transformation ever. Electric cars will be the "solar link" between renewable power and transportation energy and could provide more energy security than all our military bases in the Middle East combined. The oil companies are worried, and should be.

My friend George Perry, a former Ameriprise executive and en-ergy columnist, writes that by charging millions of electric cars at night, "We would not need to generate additional electricity for many years to come because we would capture much of the currently un-used production (about 60%) through post-midnight battery charg-ing." This scenario, not so far away, means much less polluting gasoline will be needed. Nissan CEO Carlos Ghosn, who developed the all-electric Leaf, predicts electric vehicles will comprise 10 per-cent of the automobile market by 2020. GM came back from killing the electric car to developing a production electric vehicle with en-gine assist (as opposed to the Prius and other hybrids that have an engine with electric motor assist), and it is beginning to sell well. GM's Volt was voted *Motor Trend*'s car of the year in 2011. The business opportunities for building solar charging stations for electric vehicles are immeasurable.

Electric cars are the key to energy independence, not more drilling. Elon Musk knows this, which accounts for his Tesla Motors' alliance with Honda. He envisions home solar competing with electric utilities by using car batteries for energy storage on a mass scale.

We're at an energy crossroads here, as the biggest and most important industries on the planet struggle to change. Energy and how we produce and use it is the most critical issue of our time. This book is not about predictions or what we "should" or "could" or "would" or "need to" do; it's about what we *are doing,* and what we are not.

In a little over one hundred years the world has gotten itself hooked, addicted, dependent, and entirely reliant on electricity. There are people still alive in Western countries who remember a world without it. And humans managed without it for tens of thousands of years (and as you know from reading the first ten chapters of this book, billions of people still don't have it). But don't even try to imagine the modern world without electricity! Global civilization as we know it would simply crash and millions would die, not from lack of iPads or iPhones or electric light—but because all digital operating systems would quit working, electronic communications would end, businesses would collapse, food could not be distributed, financial markets would evaporate, and quadrillions of dollars would disappear; the result would be a world as depicted in the depressing movie *The Road* or on NBC's TV series *Revolution.* Never before has humanity relied on *just one thing* to survive.

So, for this reason, electricity is the most important "thing" on the planet. We cannot live without it. We can live without oil and gas, but not without electric power. In fact, IT server farms now consume an estimated 1.3 percent of all global electricity. The amount of energy needed by the world's data centers and "the cloud" exceeds the electricity use of Australia, equal to the power from thirty nuclear-generating plants.

The good news is that we have plenty of electricity and are always making more of it, and solar is making more of it than any other *new* energy source. So this, for me, is the most exciting industry on earth. While the world depletes oil resources, quits nuclear, and shuts down dirty coal plants, we will not run out of electricity, ever, thanks to the array of solar and other renewable technologies at our disposal. For example, during a recent sunny twenty-four-hour period in Germany, distributed photovoltaics fed 22 gigawatts of solar into the grid, providing fully half the country's energy needs.

I hope this book may inspire at least a few members of the generation that will inherit this confused and chaotic planet to become energy entrepreneurs, electrical engineers, power company managers, energy financiers, solar project developers, clean-tech innovators, and inventors. The opportunities that exist right now to transform and replace the world's largest business is limitless and almost beyond the scope of our imagination.

So how will we do it?

For some answers I will turn to others whose experience, knowledge, and vision far exceeds mine, people who have inspired me for many years. I will also show what people, communities, cities, states, national governments, other countries, and businesspeople are doing around the world. It's more amazing than I could have dreamed when I started writing this book. Later, I will get to the "bad news."

One of the most esteemed researchers in photovoltaics in America is Dr. Karl Wolfgang Böer. Dr. Böer launched the Institute for Energy Conversion at the University of Delaware after leaving Germany in 1961, when the Berlin wall separated his home from Humboldt University, where he was a professor of physics. At Delaware he became the world leader in the development of the "thin film" solar cell, launched a solar company on the side, and built the first solar house in America. Constructed on the campus in 1973, "Solar One," as it was known, was heated and cooled by solar, its electricity provided by his cadmium-sulfide solar cells. The all-solar house had one hundred thousand visitors the first year—at a time when most people, myself included, didn't even know what solar energy was. At the same time, Dr. Böer cofounded the American Solar Energy Society (ASES).

I had been reading about Dr. Böer for at least twenty years but had never met him. The university instituted the Karl Böer Award for leadership in solar energy, and I was invited to the award ceremony for its first recipient, President Jimmy Carter, but I was traveling for SELCO at the time and couldn't go. Meanwhile, I was fascinated, not by Dr. Böer's academic research in solar physics, which I don't understand, but by his published pronouncements on the importance of solar energy for the future of mankind. Besides being an inventor, a professor of physics, and a businessman, he is also a poet and a philosopher. While writing this book, I decided I wanted to meet him.

I knew he'd retired and had left the University of Delaware some years ago, but I didn't know where he lived. I asked colleagues to help track him down, and then found him about five miles from my home in Florida, at a local retirement complex—serendipity again! He agreed to meet, and we spent a memorable afternoon together in March 2012. We shared stories about his friends and fellow pioneers, Hermann Scheer and Wolfgang Palz, and about the aforementioned solar scientist Richard Swanson, founder of Sunpower Corporation. He told me how his institute's research led to the thin film manufacturing process that vaulted First Solar to become, for a while, the largest PV company in the world.

I told him that the company I'd founded had just installed an 850-kilowatt PV system on the University of Delaware's field house and showed him a picture of it. He asked my help in getting a solar system installed on the new wellness center at his retirement community, which the retired Republican CEOs on the board said was a waste of money. "I don't want to live in a place without solar," he said ruefully.

His positive and well-researched vision of how solar and wind will replace fossil and nuclear fuels sooner than we think is based on a lifetime of incomparable research and knowledge. He talks, of course, in gigawatts and terawatts and says, "Worldwide, PV installations total twenty-five gigawatts, with over six gigawatts installed in the United States. That gives us over nine terawatt hours per year, enough to power 850,000 households." Noting the growth in solar beyond what anyone had imagined, he says utility-scale PV combined with decentralized residential and commercial PV, "should soon provide ten percent of our peak electric generating loads." He told me with certainty that 30 percent of our power will come from solar and wind by 2025.

Although he is a solar power scientist and visionary, Dr. Böer is also a huge believer in wind power. Two years ago, he wrote in *Solar Today* magazine, "Worldwide, wind parks have a collective total of 200 gigawatts of rated power. That's forty-six percent of all the installed nuclear power. Impressive!" There is much more today. He amazed me with the following statistic regarding the 40 gigawatts of wind power currently installed in the United States: "American wind farms are already equal in capacity to all the nuclear power plants in the United States and France together." On average, he points out, wind farms produce only 35 percent of their rated capacity. Nevertheless, he says, "American wind turbines can already supply the electrical energy for 10 million households."

I believed him. I had just driven from Indianapolis to Chicago through "Windiana" and the great Midwest wind corridor, and for one twenty-mile stretch we watched more giant windmills than I could count revolving languidly in the morning breeze above the shimmering cropland and grazing cows.

The "solar age" includes all renewables, and while this book is about solar electricity specifically, a mention must be made of the other sources of clean energy besides photovoltaics and wind. These include geothermal, large and small hydro, methane capture, tidal and ocean energy, river-flow kinetic hydropower, the aforementioned solar thermal (CSP), cellulosic ethanol, green building technologies, smart-grid energy management, biodiesel, biomass power, biofuels and algae, household wind generators, fuel cells, cogeneration, compressed-air energy storage (CAES), combined heat and power (CHP), thermal energy storage (TES), sodium-sulphur and lithium-ion batteries, advanced flywheels, and of course a host of energy-efficiency and conservation measures. I know I've missed a few. (The bibliography lists many books that cover these technologies in detail. See especially Congressman Jay Inslee and Bracken Hendricks's *Apollo's Fire* and Daniel Yergin's incomparable opus, *The Quest*.)

Today, solar and wind produce 4.72 percent of America's electricity, according to the Federal Energy Regulatory Commission; if you add hydro, biomass, geothermal, and waste heat, renewables produce 15 percent of our power.

I didn't mention hydrogen because it "isn't happening" and probably won't. At a Sierra Club "power lunch" a few years back, our old pal S. David Freeman weighed in on hydrogen: "The Bush administration has endorsed the hydrogen fuel cell, but there's no program for development of hydrogen fuel. Let's have one seat at the table for common sense, which suggests that clean technology needs a clean fuel to go with it." This is the problem with fuel cells: they require hydrogen to fuel them. Hydrogen is not an energy source but an energy carrier, and despite all the H hype, the markets never supported "the hydrogen economy." Ten years ago, the hydrogen bubble burst. One former hydrogen buff lamented as the hydrogen mania wound down in 2004, "I'm afraid that when we finally get people to stop associating hydrogen with H-bombs and the Hindenburg explosion, the next word they'll think of will be scam." Or as James Howard Kunstler, author of *The Long Emergency*, wrote in *Rolling Stone* magazine: "The hydrogen economy—widely touted as a cure-all—is a particularly cruel hoax."

I've already mentioned "breakthroughs," everyone's hope that a new technofix will solve our energy problems. Daniel Nocera, formerly at MIT's Solar Revolution department, now at Harvard, has developed his "artificial leaf," a solar-powered electrolyzer that converts sunlight to oxygen and hydrogen inexpensively, using hydrogen for storage instead of batteries. His experimental technology, he tells audiences and the press, could use hydrogen extracted from water to power millions of homes in the developing world, and eventually our own. However, as of this writing, he has no working model of a device than can convert hydrogen to electricity at a lower cost than SELCO and thousands of other small companies are already doing, using standard PV panels, cheap batteries, and LEDs.

Seven years ago, I was on a panel with Daniel at an energy

symposium in Telluride, Colorado, where he talked excitedly about this "breakthrough," and he told me later that he truly believed his "leaf" was the biggest thing since sliced silicon—that it could save the world. In mid-2012 his "transforming solar power" story was told in *The New Yorker,* but as of this writing he still had no actual operating example. We are all waiting. And waiting.

Not to entirely discount Nocera's sincere efforts, I place such academic hype in the same category as Nikola Tesla's "Free Power" device, advertised on the Internet as "big energy's secret nightmare." Using germanium diodes and electrostatic capacitors, the invention attributed to Tesla would pull "free electricity" from the earth's ionosphere. Maybe, maybe not, but I didn't send in my forty dollars to find out. Nor did I believe a Maryland inventor I knew who was building a car in his basement that ran on water, with hydrogen generated on board from the car's radio. He said everyone who had built one of these had disappeared mysteriously. He's still around last I knew. We're all waiting for the car.

Another "new" technology, this one for real, is Stirling engines, which in fact have been around for a hundred years. These were promoted a few years back as "the next big thing." One company raised tens of millions for its "Sun Catcher" technology, based on the Stirling engine, which it called "a breakthrough in solar power systems, based on proven technology and unique innovations. . . . It represents the next generation in solar power . . . ready for large-scale global deployment." I met an impressive young man with a newly minted MBA at a Clinton Global Initiative meeting who told me he'd been inspired by my earlier book to start a solar career, so he eagerly took a job with the "Sun Catcher" developer. I was honored, but I said, Why not PV? He told me their solar-dish technology was going to be bigger than PV. I had just launched Standard Solar and got worried.

In 2011, a sister company, Stirling Energy Systems, declared bankruptcy. The company's decades of work on its dream technology, which it had developed with Sandia National Laboratories, and its 1.5-megawatt demo project in Arizona "are seemingly now just a memory to be logged in the history book," as one financial analyst put it. If only that money had gone into deploying PV, investors might have been richly rewarded. I hope the young man is still in solar, somewhere; we need the dreamers and idealists.

Other darlings of the Silicon Valley investment community are Nanosolar and Konarka, start-ups that for a decade have convinced the press and investors that their amorphous PV film was the future. You will be able to paint it on walls, the promoters claim, and make it on plastic. Paul Maycock said ten years ago these technologies couldn't compete, but the nanosolar cult continues, without a viable market to date. Many of these technologies are simply distractions, contributing little or nothing to the booming clean-energy markets. While Nanosolar got another infusion of speculative capital recently, Konarka Technologies filed for Chapter 7 in 2012, and its assets were liquidated, a $150 million loss for its investors. One of its earliest investors was the state of Massachusetts; the then governor Mitt Romney personally delivered the government grant to Konarka's president—Romney's own Solyndra!

Meanwhile, polycrystalline PV panels now cost as little as sixty-five cents a watt.

As Peter Varadi told me, there have been no breakthroughs in photovoltaics in thirty years, only improvements, and Joe Romm, mentioned earlier, expanded on that when I asked him about new technology: "No technology breakthroughs in the past three decades have transformed how we use energy." My colleague Paul Maycock told me while writing this, "All that is left is crystalline PV."

There are many other fantasy technologies out there that are not going to save us: acres of orbiting space-based PV arrays that would send power down to earth via microwaves—the DOE spent $80 million on this with nothing to show for it. The Japanese are working on putting one gigawatt of PV (three square miles) into orbit twenty-five miles above the earth with lasers sending the sun power back to their island. Don't hold your breath.

Elon Musk, the rocket guy and founder of Solar City, proposes a hundred-square-mile section of Nevada or Utah could be "carpeted" with high-efficiency solar panels that would power the entire United States. If anyone could do it, he could. Meanwhile, it's more "coulds" and "woulds." We "could" do it, but we won't do it.

Then, there is fusion. Lawrence Livermore National Laboratory has been working on that for nearly half a century without success. The only working fusion reactor we know of is 93 million miles away. Sir Arthur C. Clarke once told me he believed there was "something to cold fusion," which has always sounded like science fiction to me. I checked with many energy experts, who agreed it was a nonstarter.

I mention these apparently competing technologies because almost everyone I meet is confused about solar energy, talking about the latest "new thing" in solar technology they've read about somewhere and waiting for cost-cutting breakthroughs that aren't likely to happen in time to do any good. Ohio professor, author, and energy expert Harvey Wasserman wrote in his mind-expanding book *Solartopia*, which looks back from 2030 at how we solved our energy crisis, "All the technology that was ever needed for a post-pollution world was available in 2007."

Hermann Scheer never stopped believing that we can reach "energy autonomy," supplying the world with 100 percent renewable energies with *today's* technologies that will "break the globalized chain of the fossil fuel supply networks."

A well-known PV success story is publicly traded First Solar, which, during the silicon shortage, developed a fairly efficient thin-film module using much less silicon than crystalline PV. The company, founded in Toledo, Ohio, attracted huge investment from both Wall Street and Silicon Valley. A billion dollars was thrown at it in 2008. Investors like Vinod Khosla and John Doerr, after dusting themselves off from the turn-of-the-century dot-com debacle, entered the world of clean-tech investing. Khosla became perhaps the smartest financial analyst and investor in global renewables. He now doubts thin films can compete with crystalline PV. "Most current thin-film start-up efforts do not appear differentiated enough to justify the hundreds of millions of dollars invested in this. . . . They will fail to compete in the near future," he told a Berkeley clean-tech conference.

Abound Solar, based in Loveland, Colorado, another DOE thin-film darling, using cadmium-telluride, shut its doors in mid-2012, unable to compete with Chinese crystalline PV. Its assets will be sold to pay back part of a DOE loan guarantee, fortunately not as big as Solyndra's.

One of America's greatest dreamers and doers, the late Stan Ovshinsky, a modern Edison, launched Energy Conversion Devices (ECD) and its subsidiary UniSolar thirty-five years ago. He patented the nickel-metal-hydride battery, a huge technical breakthrough, envisioning the coming electric-car market, and hired the former chairman of General Motors to run it. At UniSolar, he developed a thin-film PV product that could be rolled out on a machine the length of a football field and that was supposed to reduce solar's cost. It was flexible, lightweight, worked well in tropical heat, and you could fire a bullet through a panel without affecting its power output. Worldwide it was used mainly for commercial rooftops and "building integrated PV" (BIPV), where it could be rolled out and glued down. This thin-film PV product was also fabricated into regular solar panels, and SELF installed thousands of them in Vietnam, where they were good at charging batteries in low-light (morning and evening) and equatorial-overcast conditions. They are still there, not degrading noticeably, and working fine. Ovshinsky died in 2012.

Despite this early success and tens of millions in global investment capital, ECD and its UniSolar subsidiary declared bankruptcy in 2012. At the same time, First Solar, no longer the largest solar company on the planet, was in financial trouble despite their panels powering some of the world's largest solar farms, which they own and continue to build. Thin-film arrays take up nearly twice the space of poly and mono PV, which also now nearly matches First Solar's modules on cost. Thin film, the breakthrough technology of the century's first decade, may not make it. In 2013, GE sold its thin-film technology to First Solar and scrapped plans to build the world's largest thin-film PV factory in Colorado. Crystalline PV, the workhorse of the solar industry, is now king.

In fact, to be fair to these earnest innovators, it's not that these technologies don't necessarily work; it is that good old-fashioned crystalline photovoltaics, thanks to the unexpected and dramatic drop in crystalline solar prices due largely to China's entry into the global PV market, *beats all other PV technologies on cost.* If you had told me five years ago that this would be the case, I wouldn't have believed it. Mono- and polycrystalline PV were not only beating most other clean-power technologies, they were also killing other silicon solar technologies on price and efficiency alone.

I met with Joe Simmons, director of the Arizona Research Institute for Solar Energy. Joe believes in crystalline PV and thinks it will beat nuclear power, especially once solar energy storage becomes economic (see Appendix 1). He told me over lunch last year that "solar PV will short-circuit all other technologies, including nuclear. Nuclear is dead," he said. "It's a four-engine propeller plane being passed by jets. Once licensed, it takes ten years to build a nuclear plant, if anyone will finance it. But in ten years, PV will beat nukes hands down. It beats nuclear power now since PV is immediately deployable at any scale you want."

We talked about NuScale, the "backyard" nukes company based in Corvallis, Oregon. They claim to have solved the safety issues by burying their nuclear power modules deep underground. I'd been hearing about this for years. Like military generals, engineers and investors never give up if they think they are "right." Another "small nukes" company is Hyperion, which hypes its "nuclear batteries," reactors the size of a refrigerator—producing 25 megawatts at $100 million per copy. "Not in my backyard" nuclear power, however small and supposedly safe, will require a lot of backyards. Good luck with that.

Who is going to finance or buy a modular nuclear power plant if the equivalent amount of power can be produced by solar PV for less? Nuclear power is the most expensive electricity you can buy, double the cost of coal power, and triple the cost of wind power. Joseph Stiglitz, the Nobel Prize–winning former chief economist for the World Bank, has written, "It's questionable that without a whole set of government subsidies there would be any nuclear power plants at all."

If PV's costs can fall 70 percent in just two years, imagine where they will be ten years hence. Plentiful silicon will only get cheaper as more companies refine it, more polysilicon is produced, and PV panel production becomes entirely automated, with solar-cell factories themselves powered by PV. This was Peter Varadi's dream in 1979 when he built the "solar breeder" factory in Maryland. Thousand-megawatt PV farms—or one gigawatt—are already in the planning stages; half a gigawatt PV plants are under construction in Europe and California. A 1-gigawatt solar farm, covering ten thousand acres, equals a 1-gigawatt nuke plant, and there are 104 nuclear plants in America, all over thirty years old, that will soon need replacing or decommissioning at huge cost. One-gigawatt solar farms are cheaper to build than 1-gigawatt nuclear power plants, *today*.

Meanwhile, while the cost of solar plunges, the cost of fossil fuels can only go up, except for natural gas. Thanks to shale gas "fracking," gas prices have fallen 80 percent in the past couple of years, which will certainly be a setback to the rush for renewables, and even nuclear. But cheap gas may only be temporary, especially if we begin to export it to Europe and Asia, where prices are five times higher than here. Prices could shoot back up. *Rolling Stone* magazine writer Jeff Goodell (author of *Big Coal*) claimed in 2012 that America's largest shale-gas company, Chesapeake Energy, had engaged in a "scam" that was responsible for the "big fracking bubble" that could eventually collapse, resulting in unexpected price fluctuations. My friend John Ewen, an investment banker at Ardour Capital in New York, who encouraged me to start Standard Solar but couldn't convince his firm to invest, says the appeal of solar power as a long-term investment is that the sun's cost as a fuel is predictable and essentially free. "It's clearly a better credit to bet on the sun coming up tomorrow than the price of natural gas," he told *The New York Times*.

However, there is one ray of ancient sunshine: thanks to the increased use of natural gas by our electric utilities—replacing coal—America's CO_2 emissions have dropped 20 percent! According to writer Michael Clare, America's unexpected energy gold rush has ushered in the "Age of Unconventional Oil and Gas—the third great carbon era." Oil prices today opened an opportunity for expensive new technologies, originally developed by the DOE, to extract oil and gas out of shale rock formations deep in the earth. We've cut oil imports, and politicians tout our new "energy independence."

Environmentalists, on the other hand, say if we don't leave most of these fossil fuels in the ground, the Earth is doomed. We can become energy independent without burning carbon; it's a political decision.

You can't talk about solar without talking about gas and nuclear power. Cheap gas hurts solar, temporarily, but it really kills nuclear. Gas can replace coal, and it is doing so, but it also generates greenhouse gases, even if only half as much. Nuclear generates no carbon emissions (other than during plant construction), so even some environmentalists are giving it a second look. James Lovelock, the originator of the "gaia theory" concerning the role of the thin membrane of enveloping atmosphere that provides for all life on earth, argued in his book *The Revenge of Gaia* that whatever the risks posed by nuclear power, it was a safer bet for the world than the dangers of climate change caused by continuing to burn fossil fuels (he wrote the book before it was clear that solar could take up the slack).

For years, some thought, myself included, that *maybe* we needed to learn to live with nuclear energy so it could provide "base-load power" since solar and wind are intermittent. Solar only makes hay while the sun shines. This drives utility managers, who are mandated to buy a certain amount of solar by state renewable portfolio standards, nuts, requiring them to build equivalent conventional capacity as backup. (See Appendix 1: "Solar Tech Simplified," for a detailed discussion of energy storage.)

Then, along with a huge drop in solar prices, we suddenly had an 80 percent reduction in natural gas prices! Here was a new base-load energy source—combined-cycle gas turbines that love to idle and can come online instantly when the sun quits shining. But this didn't stop the nuclear guys who don't care about cost or risk since they can get the government and ratepayers to cover them. And France uses it, people point out, so why can't we? Because the French government paid for it. Will ours?

Then, to add to the confusion about our energy future, along came Japan's Fukushima catastrophe! Just when we'd begun to forget about Chernobyl and Three Mile Island, here was what many think was the final nail in the nuclear coffin. It just may be. As one

Fukushima city resident said, "We should not build things that human beings can't control."

I questioned Joe Simmons as to why the DOE's Secretary Chu had approved an $850 million loan guarantee to two recently licensed nuclear plants in Georgia, and why two more were about to be licensed in South Carolina. "The DOE are idiots," he said. "Steven Chu was a big backer of nuclear before he turned to solar," he noted. Meanwhile, many "liberal greens" see the "new nukes"—smaller and safer—as the best non-polluting source of power.

The DOE, employing twelve thousand scientists and seventeen national laboratories, has over the years made many substantial contributions to America's energy mix, especially in solar. However, loan guarantees that skew the market, encourage the P. T. Barnums, and often back the wrong technologies is not one of them. The DOE's Energy Loan Program, funded by President Obama's 2009 stimulus package, has invested in several large photovoltaic and solar-thermal utility-scale power plants, and these are a success since there is a guaranteed return on investment. Investing in the Tesla electric car is a success story. There are probably other Solyndras out there amid the batch of forty projects the DOE put money into; there is no room here to discuss all the successes and failures of the department since it was launched by President Carter.

However, I can say this: in the future, the DOE needs to stay out of solar investment and loan guarantees and instead focus on energy policy, utility regulation, electronic power management, smart grid technologies, data collection, materials innovation, advanced clean-tech research, power-market integration, tax policy, and incentive grants. Let the markets rule, not the government, and they will rule in favor of workable and practical and affordable clean-energy technology. It's been suggested that the DOE's loan program be replaced with a "green energy development administration." Well, maybe. Asked recently how government's can emulate Silicon Valley, Vinod Khosla said, "Stay out of our way."

As for spending too much on alternative energy, have a look at the facts. The United States is spending $3 billion a year on green-energy research and $77 billion a year on defense R&D. In any case, the stimulus subsidies are gone, and the 30 percent solar tax credit expires in 2016. "Solar has enjoyed subsidies for twenty years," says Paul Maycock, whom I trust more than anyone for insightful analysis of the solar industry. "We don't need them anymore. Without subsidies, PV is fully economic now in 30 percent of the world." *For Americans, solar electricity without subsidies at all will be competitive with the grid by 2018 and in many U.S. markets grid-parity is here now.*

Meanwhile, Secretary Chu, at President Obama's behest, pushed stimulus money into renewable energy, adding, according to a government report, 13,000 megawatts of solar and wind capacity to the grid, creating 75,000 new jobs while generating $25 billion in economic activity. Except for Solyndra, almost all of the stimulus money for clean energy was largely well spent and helped with the country's economic recovery, creating tens of thousands of jobs. Standard Solar saw its share of stimulus money via Treasury grants that helped level the playing field for solar to compete with subsidized fossil fuels, an excellent program. We never asked for nor wanted loan guarantees. Anyway, to put it in perspective: the burden to taxpayers of Solyndra's bankruptcy is equivalent to what we were spending in 2012 in Afghanistan every two days. Congress never complained about that.

The day before Dr. Chu was sworn in as President Obama's secretary of energy, he visited a Standard Solar installation on a D.C. elementary school. Tony Clifford and I showed him around, and he asked me what I thought of thin-film PV. I said it will never compete with crystalline solar cells, wasn't as efficient, degraded over time, took up too much space, and could never be used on rooftops like the one he was looking at that morning. However, Dr. Chu, when he decided to push solar at the DOE, chose to back thin films, a big mistake as the bankruptcies of the thin-film companies Solyndra and Abound Solar would later illustrate.

I had never spoken to a Nobel Prize winner before! Dr. Chu was very nice, humble, smart, but it was clear he didn't know solar. I did tell him to be sure to take the stairwell next to what would be his office on the top floor of the DOE's Forrestal Building and go up on the roof and see the 210-kilowatt solar array we had installed. He had no idea it was there. The following evening at the Environmental Inaugural Ball, I wanted to follow up with him on the subject of solar, but he was surrounded by his new Secret Service detail, and Tony and I couldn't get near him.

America needs to learn from Germany. I've already talked about my friend, the late Hermann Scheer. A Social Democrat member of the German parliament for twenty-five years, the former student antiwar activist and "68er" is largely responsible for launching the German solar revolution. Scheer and his Social Democrats pushed through the first-of-its-kind energy pricing innovation, the so-called feed-in tariff, which first became law in Germany in 1991. It was inspired by America's Public Utilities Regulatory Policies Act (PURPA), which was passed by Congress in 1978 during the Carter administration.

The feed-in tariff, or FIT, pays for power fed into the electric grid by homeowners and businesses at a guaranteed price, set by the government, that utilities must pay for renewable energy, originally only hydro and wind power. In 2000, Scheer shepherded into law the German Renewable Energy Act that expanded FITs to include PV at rates much higher than conventional electricity. The money comes from the people, who in Germany voted for this; all ratepayers have the price of a cup of coffee per month added to their monthly electric bills to cover the cost of the FITs.

It was a long and difficult battle to get such aggressive renewable energy legislation passed in Germany; the utilities fought back hard. Scheer, the consummate Bundestag politician from Germany's largest state, North Rhine–Westphalia, gave up on the government taking any action, despite its commitment to combat climate change. He told me, "There is no free market in energy, and governments are the last to move. So we relied on the people. We went to the people with the initiatives, who voted in favor of solar electricity."

He proved that people were happy to pay more for clean, distributed, locally owned sun power. Today, nearly a million solarelectric systems cover German rooftops.

Can you imagine this happening in America? Republican Party leaders, who mostly don't like solar anyway for inexplicable reasons, would consider a rate increase of two dollars a month an intolerable tax burden and would kill it. Unlike Germany, which has a handful of utilities, we have nearly five hundred, as well as fifty state-public-utility commissions, making widescale change like this nearly impossible. However, local communities, including Gainesville, Florida, and San Diego and Palo Alto, California, have instituted pilot FITs, copying Germany. In 2013 the Los Angeles Department of Water and Power put up $150 million to buy solar power at above-retail rates. More municipal FITs are on the way with support of many publicly owned power companies.

In Germany, thanks to Scheer's tenacity and vision, along with the people's will, solar electricity was purchased by utilities from homeowners and businesses for triple the retail price charged ratepayers. Solar power exploded overnight, quickly surpassing Japan, the world's leader until then. Anyone with a viable roof could get a solar system financed easily, and everyone made money: the banks, homeowners, commercial building managers, "solar park" developers, the German solar-manufacturing industry, installers, and electricians. It cost the government nothing. Eventually, the artificial price was reduced year by year, but 50 percent of all the solar PV installed in the world as of 2012 is still installed in Germany—7,000 megawatts a year for the past two years—in a country that you don't think of as exactly sunny. Germany has 25 gigawatts of installed PV capacity.

Germany has slashed its FIT subsidies to rein in its solar boom. Today, the FITs could be dispensed with since grid parity has been reached in Germany, thanks to the fall in PV prices. Mission accomplished! And now, as I've already mentioned, grid parity is becoming a reality in many energy markets, including parts of the United States, sooner than anyone predicted. In Germany today the utilities are losing money on their once-profitable coal-fired peak-load power plants as solar cuts peak pricing by 40 percent! In California, solar PV is also beating peak power rates, causing pushback by seriously unhappy investor-owned utilities.

After the Fukishima disaster, the German parliament made a nearly unanimous decision to close down its seventeen nuclear reactors, relying on solar to make up part of the difference. Unfortunately, this also means more electricity from coal, since solar and wind can't yet cover the loss of nuclear power. As much as Germans fear global warming caused by fossil fuels, they fear nuclear power more—having been geographically too close for comfort to Chernobyl. In any case, Germany will continue to be the crucible for sane, viable, clean energy policies. "The German energy transformation is as challenging as the first moon landing," said the CEO of Germany's second-largest utility. "It's a huge challenge we'll be able to master only if everyone works together."

Flying into Munich in 2012, I looked out the window and was dumbstruck by what I saw below on the glide path over the rural landscape: nearly every barn and most farmhouses were covered in PV! Factories with PV then came into view. And driving through southern Bavaria, I saw large rooftop-PV systems everywhere I turned, and it was rare to see a barn without solar. Cruising down the Danube, one is treated to a rhapsody of solar homes sparkling among the riverside villages.

Spain and Italy are the next biggest user of photovoltaics; both instituted FITs, following Germany.

Italy is the second-largest market for PV; its solar installations grew 245 percent in five years, overtaking Germany in 2011. Nine gigawatts of solar PV were added to its grid, supplying 2.5 percent of Italy's power. PV is everywhere, as I witnessed during a recent summer, covering Autostrada rest-stop parking lots, used for half-mile-long sound barriers along urban highways, sprouting on tiled rooftops in every village and town. Giant solar farms supply the nation's utilities. The European Photovoltaic Industry Association (EPIA) originally predicted solar PV in Italy would be competitive with conventional grid power by 2013, but the country reached grid parity in 2011. Sun power surpassed wind power in Italy in 2012. ENL, Italy's giant utility, is now franchising solar installers—750 to date. This is a model that U.S. power companies might look to one day.

Spain has a long history in photovoltaics. Peter Varadi's Solarex was installing off-grid solar home systems in remote villages in the Pyrenees in the 1970s. Besides a massive residential PV program, two of the largest concentrated solar power plants in the world are in Spain, as are some of the biggest PV solar farms. One solar thermal plant between Seville and Cordoba has 2,600 solar mirrors focused on a towering heat exchanger, where molten salt produces steam to run turbines and energy is stored to produce power twenty-four hours a day. The Spanish firm Abengoa Solar has built a similar 753-megawatt solar concentrator plant in the United States and has started work on another one.

In the United Kingdom, solar took off after my old Greenpeace colleague, Jeremy Leggett, who helped me raise funds for SELCO in the midnineties, founded Solar Century. His company, in 2012, did over $100 million in revenues, thanks in part to Britain's FIT. Solar grew dramatically in Britain over the past decade. For example, Sainsbury's installed 16 megawatts of PV on its 169 grocery stores, becoming Europe's largest solar generator. But solar slowed down under a conservative government that decided to cut funding for feed-in tariffs. "The utilities are viciously fighting solar with a ferocity I never expected, and now they have the ear of the government," he told me on the phone from London. I asked him if this meant the politicians no longer backed solar solutions to climate change. "Not at all. It's about profits. The utilities don't like solar competing with them. No politician here or in Europe would dare deny climate change like they do in the U.S. They'd be laughed out of office."

While the United Kingdom backslides, Europe continues to light the way with solar PV, with FITs in place in many countries and over two and a half million PV installations. France, where we all know nuclear power is still the dominant energy source, is nonetheless expected to surpass California in PV installations, mostly residential, in 2012. As I mentioned, France's largest oil company bought America's second-largest solar company, Sunpower, in 2011. And a French company is building a 44-megawatt PV power plant in Puerto Rico.

Portugal for a time boasted the largest PV array in Europe, surpassed by a 100-megawatt PV plant in the Ukrainian Crimea, which topped the 90-megawatt farm south of Berlin.

The Netherlands, which has encouraged residential rooftop solar for twenty years through generous subsidies, recently announced a strategy to insure installation of between one and two million residential PV systems by 2023.

Sweden, celebrating Europe's first "Green Capital," plans to reduce per capita carbon emissions to 1.5 tons annually, compared to the current twenty tons a year for every American.

Belgium roofed 2.2 miles of railway tracks with PV to help power its electric trains.

Denmark, the world's leader in wind power, is committed to 100 percent renewable energy by 2050—the country's energy policy has resulted in a 21 percent decrease in carbon emissions while its economy grew by two thirds since 1980.

Greece, with only several hundred megawatts of PV installed to date, is seriously looking at solar to create economic prosperity—they want to export sun power to their northern neighbors.

Bulgaria brought its 60-megawatt Karadzhalavo PV plant online in mid-2012, built by America's SunEdison, which is now working in twenty-nine countries. It was built in just four months.

Thailand plans an additional gigawatt of solar with new feed in tariffs.

Exporting solar power from sunny southern countries to northern Europe is not a new idea. Consider the heralded Desertec project, which will either be the grandest solar solution ever or a colossal solar boondoggle. In 2009 a consortium of governments, utilities, and investment banks proposed to turn hundreds of square miles of the Moroccan desert into a vast CSP array to send power northward on a supergrid through low-loss transmission lines under the Mediterranean Sea. However, construction on the first 500 megawatts of concentrated solar power has been delayed. Like many of the other solar-cum-pie-in-the-sky technological fantasies, this one is the biggest to date—it will cost at least $400 billion, the largest industrial investment in history. It could work, but will it get built?

Dr. Scheer discounted the "premature euphoria" of Desertec in his last, posthumously published book, where his devastating disquisition is far too long to present in these pages. Suffice it to say he believed this "new megalomania" will only divert money and attention away from distributed solar power for the people and toward massive centralized ownership of renewables by unaccountable corporations and global financial elites. He also shows that Desertec and supergrids will be more expensive than decentralized, distributed solar and wind in the long run, not to mention that there isn't enough water in Morocco to supply CSP on this scale. When Desertec was in its early idea stages, Chinese solar panels had not yet turned the global solar industry upside down.

To wit, in a very different approach to the direct sourcing of PV product, DHL, the world's largest logistics company with over $50 billion in revenues and 275,000 employees, set up a distribution center for Chinese solar modules in the Netherlands in 2012. Thanks to China, for better or worse, solar PV has become a commodity, even with the 30 percent import duty the United States slapped on Chinese modules in May 2012. Wal-Mart will be selling solar next, and people will order their solar home systems online to be drop-shipped for local electricians to install. PV Depot was retailing 290-Wp Chinese-made panels for $0.89 a watt. China itself will have surpassed Germany in domestic solar installations by the time you read this.

In the rest of the world: Taiwan built the first 100 percent solar-powered stadium, covering fourteen thousand square feet of roof space. In Brazil, seven of the twelve venues for the 2014 World Cup are being powered by rooftop solar; Qatar is planning a state-of-the-art solar-powered stadium for its 2022 World Cup games; Abu Dhabi next door is building Masdar, an all-solar city of forty thousand with a $22 billion price tag. Saudi Arabia announced plans for 54,000 megawatts of renewable energy, including 16,000 megawatts of PV and 25,000 megawatts of CSP and will pay for it with feed-in tariffs. Given the current price of oil, it's cheaper for Arabia to produce electricity from the sun than to burn valuable, exportable oil. The Arabs also know something we apparently don't; oil won't be there forever.

Japan never had oil, so it went into solar in a big way in the early 1990s. Their solar factories, long before China got into the business seriously, supplied half the world: Kyocera, Sharp, Mitsubishi, and Panasonic (which bought Sanyo) continue to rank among the largest PV manufacturers in the world. Sharp, Kyocera, and Mitsubishi opened factories in the United States. Following the Fukushima disaster, Japan is resetting targets for solar and turning back to renewables. Toshiba announced in 2012 that it will begin building solar plants with a total generating capacity of 100 megawatts on the country's tsunami-wrecked northeast coast. Japan, again, is poised to become the world's hottest solar market. Japan's chief competitor, Korea, is getting into PV, and Samsung Solar, Hyundai Solar, and LG are delivering 19 percent efficient solar panels to the world markets.

South Africa, with some of the highest solar radiation on the globe, has awarded contracts for 632 megawatts of PV, 150 megawatts of CSP, and over 600 megawatts of wind, even while they have more coal than they know what to do with. They also have common sense. As host to the Durban Climate Change Conference in 2011, theirs is a government that does not want to contribute more to global warming than they have to. In nearby sun-saturated Namibia, an American investment group has secured a power purchase agreement from the government to sell it 500 megawatts of distributed solar power, sized from rooftop to 20 megawatt installations, replacing diesel generators. Jigar Shah, one of the project's partners, is going after the world's diesel generator market with solar PV. "Ten percent of the world's electricity is supplied by diesel today," he told me, "easily replaced by solar at a big cost savings." His solar projects expect a 12 percent return on investment.

Sunny Australia's six hundred PV dealers are selling oodles of solar. One, called True Value Solar, installed 12 megawatts in one month not long ago. Another Aussie firm, CBD, bought Westinghouse Solar in the United States to get into the American market. Tim Flannery, author of *The Weather Makers,* chaired a climate change commission in the Australian state of Victoria that claims the state's solar potential would allow it to get all its electricity from the sun.

Islands nations, like the Maldives in the Indian Ocean and Tokelau in the Pacific are striving to become 100 percent solar powered. Many years ago, I met with the energy minister in the Maldives and the manager of their diesel-powered generating plant to try to talk them into going solar. Fifty percent of their foreign exchange was going to buy oil to run generators in the capital and throughout their hundreds of islands. They didn't buy the economic argument, believing solar wouldn't be practical or affordable. Years later, once they began to witness the result of sea-level rise that threatened to sink their nation, they decided it was time to be an example to the rest of the world and put the sun to work. In effect, they are saying to us that our folly is their doom: "If we are willing to go solar, why won't you?"

Closer to home, Canada is going solar. Ontario passed North America's first feed-in tariff several years back, resulting in the installation of nearly 400 megawatts of solar PV by 2012. It's mostly residential, but they also built an 80-megawatt PV plant with First Solar thin-film panels, the largest solar farm in the Western Hemisphere at the time. It can power 12,800 homes. Not exactly a sunny place, Ontario is expected to surpass California in solar installations, even while they reduce their ridiculously generous FIT.

So what are we doing? Plenty, but we could do much more. There is no excuse for the U.S. solar market to fall behind Italy and Spain considering that we invented the modern solar cell, launched the photovoltaic manufacturing industry, and are five times bigger. China will soon surpass us; it was once a country so poor that when I visited peasant communes in the 1980s, farmers were proud to show off their prized possessions: a bicycle, an alarm clock, and a new thermos bottle. Now, half our solar panels are made in China. It is becoming the new Empire of the Sun, while we slip backward.

Politics is the reason (here comes the bad news). You can't talk about energy without talking about politics. Oil dominates our political world, but it is electricity that is the most political, since utilities and our power networks are government regulated, be they investor or publicly owned. Solar electricity feeds power into the grid and thus becomes part of the public domain. This book won't attempt to sort out the sorry state of American politics in the twenty-first century, but I will outline the impact our sclerotic democracy is having on energy policy.

First, we don't have one. Because the people demanded it, we do have a patchwork of laws, regulations, tax breaks, and incentives for clean energy going back to the Carter days. But today the people are no longer in control. Special interests are, and the American people are no longer a special interest. Vested, wealthy, minority interests refuse to cede power, figuratively and literally.

However, according to an ABC/*Washington Post* poll, 92 percent of the American people favor pursuing solar (and 83 percent favor wind, as well). A recent Kelton Research poll found that 9 out of 10 Americans support the use and development of solar technology. Great, so why aren't we doing it more than we are? Why are we ceding our energy future to China, Europe . . . and Saudi Arabia? And to the Koch brothers?

The answer is simple. We are not pursuing clean tech like other countries are for the same reasons that we are not rebuilding our roads, repairing our bridges, upgrading our electrical grid, building high-speed rail, fixing our educational system, and providing health care to our citizens (like every other industrial democracy). This is "American exceptionalism" at its worst.

Three simple things now determine our political life: (1) the once lunatic fringe of American politics has gone mainstream because of rage over wars, spending, and social issues; (2) the Republicans are misusing an arcane, undemocratic rule of the U.S. Senate requiring sixty votes to pass anything; and (3) the Supreme Court decision (*Citizens United*) giving corporations and billionaires the right to finance political campaigns with unlimited funds, killing campaign reform. (Even Senator John McCain, supporter of overhauling campaign finance, said the Citizens United ruling was the dumbest decision in the history of the court; TV commentator Chris Matthews said it may have killed our democracy.) The oil lobby spent $270 million on TV ads attempting to defeat Obama.

Thus, we have lately been ruled by a half-crazed, misinformed, and paranoid minority whose ideological lunacy controls our second political party and now rules our Congress with unlimited funds. Majority rule is dead at the federal level in America, and so therefore is democracy, which is defined as rule by the majority. The ability of the American people to determine their own destiny, which would include, as all polls point out, a clean-energy future, is in grave doubt.

Powerful oil, financial industry, and war-profiteering elites are in control, getting richer as they go. If you don't believe me, read *It's Even Worse Than It Looks: How the American Constitutional System Collided with the New Politics of Extremism,* by political writer Thomas Mann and veteran pundit Norman Ornstein, published in early 2012.

It's the most divisive political climate I've ever seen since watching President Eisenhower stump for Richard Nixon in Cleveland's Public Square when I was in high school and listening to Senator John Kennedy make a campaign speech at our town's park bandstand. I've forever after followed politics closely, and recall when the laughable sociopaths of the John Birch Society met in church basements, funded by people like the Koch brothers' father. Today, the billionaire brothers fund Americans for Prosperity and Vote4Energy.com, which run endless TV and Web ads denouncing the Obama administration's support for "green energy." They and their "greedy bastard" compatriots' money back every extreme-right candidate running for governor, Congress, or the presidency. Read the book *Greedy Bastards,* by MSNBC's financial commentator Dylan Ratigan.

Meanwhile, our media is controlled by a handful of powerful corporations, pushing independent voices to the sidelines. As a result, moronic radio commentators and unbalanced TV news groups spread misinformation and disinformation to millions, literally brainwashing enough citizens to seriously impact political debate. Roger Ailes, perhaps the most influential and dangerous man in America, is feared, reportedly, even by his boss, Rupert Murdoch.

You no longer can talk about climate change and global warming in political campaigns, and the mainstream media no longer discusses an issue that was relentlessly covered by every American magazine and that dominated electronic media and civil public discourse less than a decade ago. A powerful, misinformed minority now despise environmentalists and refuse to believe the 97 percent of scientists who say humans are warming the planet. Elected fools confront environmentalists who want to reduce the 36 billion tons of greenhouse gases pumped into the atmosphere every year. Because environmentalists, their enemy, are for clean energy, then clean energy must be bad.

The energy-security argument and "getting off foreign oil" got lost along the way. Even the message of former right-wing political financier and oilman T. Boone Pickens, that we must stop sending $2 billion a day to the Mideast, is ignored. Pickens has built a 377-megawatt windfarm in Texas, putting his money where his mouth is. Rather than spend billions on clean energy, America chose to spend over a trillion dollars keeping the sea lanes open for the oil industry, fearful of a bunch of backward, ragtag Arab extremists.

Al Gore, who attempted to enlighten the world about the threat of global warming and who proposed taxing carbon in his 1990 book *Earth in the Balance*, is today a laughingstock to some of America's major media. At Constitution Hall in Washington in 2008, before a packed house, Gore launched his campaign for America to generate 10 percent of our electricity from solar and wind in ten years: "ten in ten" was the slogan. The media ignored him. If you want to keep your job in TV, where most Americans get their news, don't mention global warming, even when reporting that 2012 was the hottest year ever recorded, breaking fifteen thousand records worldwide—in the United States temperatures averaged 2.8 degrees higher than the twentieth-century average. A majority of TV meteorologists now deny evidence that warming is affecting weather. And don't push the huge success of solar and wind in this country, only the odd failures. Strange that in the "age of information" we may have the most uninformed and *mis*informed electorate of any Western democracy. (Part of the reason is the lethal virus of twisted lies that circulate on the Internet as "facts.")

Even stranger is that while nine out of ten Americans *want* more sun-powered electricity—including 84 percent of Republicans, according to a recent poll—a powerful, misguided minority continue to elect fools (the only word for many Tea Party Republicans) to public office who block every legislative effort in support of solar while they advocate increased defense spending. Beholden to their plutocrat supporters, they killed "cap and trade," which would have put a price on carbon emissions, and proudly defeated a Democratic bill to end oil subsidies running $4 billion a year. Cognitive dissonance be damned; stupidity rules. And today it's far worse than

2004, when Britain's conservative *Daily Mirror*, announcing the re-election of George W. Bush, headlined on its front page, "How Can 59,054,087 People Be So DUMB?"

Perhaps it is because we are living in what the late actor Pete Postlethwaite called "the age of stupid," also the title of his 2009 feature film that looked back at a planet destroyed because we ignored the warning signs of climate change and failed to implement available solutions. His movie received worldwide distribution and acclaim, but it had no effect. No TV station in this country will ever show it.

We all know what happened in Germany when a minority of angry, misled, and confused citizens took over a democratically elected government. One day, Jews were upstanding citizens and part of public life, leaders in business and the arts, and the next day they were the enemy. Lies work, as the German propaganda ministry learned. We are learning this, too, as imbecillic radio and TV personalities dominate the airwaves, brainwashing a now-influential American minority that is up for grabs by demagogues.

Ben Bova, one of America's most successful science fiction writers with 124 books to his credit, also thinks a lot about real science. In his newspaper column, observing how "the ignorant march by endlessly, toward oblivion," denying science, he wrote, "there are the people who don't accept the idea that our global climate is changing. . . . They fund studies and advertising campaigns that claim that greenhouse warming is at best a misreading of the evidence, and at worst a hoax." A cautious pessimist himself, Bova concludes, "We will continue to burn coal and oil and natural gas. We will continue to hold meetings and issue warnings. Until that tipping point is reached. Then we will die. By the tens of millions."

Ben urged me to write this book because he imagines—as he envisions interplanetary travel by our kind—that humans, in spite of ourselves, can do better.

James Hansen, the former outspoken NASA Goddard scientist, continues to sound the apocalyptic alarm on climate change, writing in *The New York Times* in May 2012 that civilization is at risk if we continue to do nothing about fossil fuels, especially Canada's tar sands. If Canada exploits its tar-sands reserves, and if the XL pipeline is built, he wrote, "It will be game over for the climate." He added, "Instead of placing a rising fee on carbon emissions to make fossil fuels pay their true cost, leveling the energy playing field, the world's governments are forcing the public to subsidize fossil fuels with hundreds of billions of dollars per year. This encourages a frantic stampede to extract every fossil fuel through mountaintop removal, longwall mining, hydraulic fracturing, tar sands and shale extraction, and keep ocean and Arctic drilling."

Meanwhile, by 2013 CO_2 in Earth's atmosphere had reached 400 parts per million, the highest known in eight hundred thousand years, based on ice-core samples. As 2012 was breaking all heat records, the Department of Defense determined fossil fuels were "a national security threat," that climate change was a "threat multiplier." Thanks to extreme weather, Americans suffered more power outages than ever before: "We've got the 'storm of the century' every year now," said a senior vice president and a thirty-eight-year veteran of PEPCO in Washington, D.C. This was a year before superstorm Sandy devastated the East Coast. Ignorant ideologues in our fair land continue to deny humans have created this increase in heat-trapping gases, and the nation remains more divided on this issue than ever. Fossil fuels rule.

We could pass a simple carbon tax, but we won't. The revenues would revolutionize our economy, create millions of jobs, support clean energy, subsidize the working class impacted by rising energy costs, finance public transport, and rebuild our deteriorating infrastructure. Historians will not be kind to the people who stood in the way of this clear opportunity to save America. At least we can take heart that President Obama made tackling climate change a central vow in his second inaugural speech. In July 2013, he announced his ambitious Climate Action Plan to cut greenhouse gases and use the Clean Air Act to reduce coal-plant emissions. Senator Mitch McConnell immediately accused Obama of launching a "War on Coal." It's about time!

Anyway, this is not a political book, and I don't know why our political system is broken or why humans can be so stupid, but I had to honestly answer the oft-raised question, "Why aren't we using solar?"

* * *

Let me get back to the good news. *We are using more solar than most Americans are aware of.* This positive story goes largely untold because the mainstream media no longer celebrates clean tech, even while President Obama dared to extol his green-energy programs during his presidential campaign. One hundred twenty thousand American workers are now employed in solar, increasing by twenty thousand a year. America's top-ten utilities integrated 561 megawatts of solar electricity capacity in 2010, a 100 percent increase in one year. And while the federal government has achieved enormous success in fostering renewable energy over the past two decades, it is now the states and cities that are making the biggest difference, responding to the people's will to address pollution and climate change. Because states have implemented renewable energy, 151 planned coal-fired power plants have not been built, and 50,000 megawatts of coal plants have been retired. In 2011, President Clinton and New York mayor Michael Bloomberg merged their climate change initiatives to focus on America's—and the world's—largest cities, where they believed effective action would get the quickest results.

The states and cities are justifying Amory Lovins's observation that "In our immensely diverse and politically fractured society, renewables can thrive if we focus on outcomes, not motives—if we simply do what makes sense and makes money, without having to agree on why it's important. . . . You don't have to believe in climate change to solve it."

This parallels the story of the Prius, which people bought in large numbers in order to be green and not pollute; today, millions buy 52 MPG Prii and other hybrids to reduce their "pain at the pump." Not polluting is a side benefit but no longer the market driver.

It's not just about the environment anymore; solar makes money— maybe not through manufacturing, since it is struggling with Chinese competition, but through sales, installation, and solar electricity production, which are going gangbusters, as I can personally attest, having seen the company I started prosper right through the longest recession in our history. It is at the state level where these truths are manifold.

I can't cover the solar revolution in all fifty states, but here are some highlights:

California, of course, started it all with the establishment of its forward-looking California Energy Commission decades ago. The California Energy Plan of 2005 called for 33 percent renewable energy by 2020, and the state's "Million Solar Roof Program," promoted by Governor Arnold Schwarzenegger, a Republican, made California number one in solar power. He now pleads for sensible energy policy from fellow Republicans, writing in *The Washington Post* in 2011:

> It is absurd that our federal government spends tens of billions of dollars annually subsidizing the oil industry . . . while the industry focused above ground on wind, solar and other renewable energy is derided in Washington. If the candidates running for president believe in energy independence as a matter of national security—regardless of whether they agree with the science behind climate change—then the issue of investing in renewable energies must be front and center in the campaign.

He noted that "our green sector is the brightest spot in California's economy, having grown 10 times faster than any other business sectors since 2005."

I used to say to my European friends, "Germany is seven years ahead of California in solar, and California is seven years ahead of the United States." State government, both political parties, utilities, cities, and communities work together in the Golden State, resulting in solar becoming cheaper than natural gas as hundreds of megawatts of large central solar plants come online, producing 567 gigawatt-hours of electricity. In Antelope Valley, Warren Buffet's MidAmerican Solar and SunPower Corp. began construction of two PV megaprojects totalling 579 megawatts. Another 130-megawatt solar plant was financed by Tenaska, the giant natural gas firm.

But it's not just utility-scale solar. There are 3,500 green-energy firms in California, employing over twenty-five thousand people. Many of these are the installation companies that put solar on thousands of residential rooftops, outfits like Solar City, which started in the Bay Area and has since gone national, landing the $350 million "Solar Strong" contract to supply solar to 120,000 houses on U.S. military bases (Solar City now competes with my former company, Standard Solar). Solar City raised $1.4 billion in 2012, including $280 million from Google, to finance its residential power purchase agreements (PPAs). Earlier, I mentioned their $500 million fund from Goldman Sachs to finance residential PV. The company closed a $92 million IPO in late 2012.

In 2013, JP Morgan Chase invested $630 million in Sunrun's rooftop solar fund. Thanks to neighborhood purchases of residential solar systems, grid parity has been reached in parts of Los Angeles. A California nonprofit, Grid Alternatives, working with China's Yingli Solar, is providing solar systems for four hundred low-income families. Twenty percent of all new homes in California will include solar systems.

Lancaster, in California's high desert, aims to be the nation's "solar capital." It has thirty-nine megawatts of solar PV installed on businesses, homes, schools, civic buildings, and parking lots and is adding fifty megawatts more. "We want to be the first city that produces more electricity from solar energy than we consume," says Lancaster's Republican mayor, who sees solar as a job producer. Getting a permit for a solar installation in his town takes fifteen minutes.

Last year, Palo Alto purchased 80 megawatts of solar at 6.9 cents per kilowatt-hour over thirty years, cheaper than the conventional power the city buys.

In San Francisco, David Llorens's company, One Block off the Grid (1BOG)—silly name but an innovative firm—put the Groupon model to work for solar, aggregating customers and procuring cheaper solar installations for communities. Llorens believes the market for solar in the U.S. is "gigantic . . . it's just enormous, and it grows at breakneck speed every year." He points out that the tipping point is here now, as it was once for air-conditioning that went from window units to central air almost overnight.

As an aside, regarding air-conditioning, let me add this thought: whoever invents a simple, solar-powered AC package that runs superefficient compressors and fans on direct current could become one of the richest people on earth. Many have tried, but there is still no affordable commercial PV-powered air conditioner being marketed as a global commodity. One reason for India's huge blackout was the dramatic increase in air-conditioning. Ironically, while it will be a good thing when solar grows as fast as central air did in the United States, it will be a very bad thing if air-conditioning in the two-thirds world grows as fast as the worldwide solar business! In this case, massive electricity blackouts will become the norm, unless nations decide to also go solar. The marriage of AC and solar in the developing world is perfect in that it's hottest in the daytime when the sun shines the brightest.

In the West, meanwhile, solar power beats the grid during peak demand wherever there is time-of-use metering and "tiered pricing," which California pioneered many years ago. Nationally, the DOE's SunShot Initiative aims to reduce the cost of residential solar electric installations by 75 percent before 2020. GTM Research reported in August 2013 that in the United States one solar photovoltaic system was being installed every four minutes! "Solar is growing so fast it is going to overtake everything. It could double every two years," said Jon Willingham, chairman of the Federal Energy Regulatory Commission (FERC). Rhone Resch, president of the Solar Energy Industry Association, says, "By 2016, solar electricity will be the lowest-cost source of retail power in the U.S."

Independent solar contractors are installing solar for America's largest home builders, such as Lennar and D. R. Horton, and putting it on the rooftops of 150 Wal-Mart stores, turning variable costs into fixed costs (Wal-Mart expects to have 1,000 solar-powered locations by 2020). Dozens of government agencies and thousands of commercial properties and factories have become part of California-encouraged *distributed generation*. Bank of America Merrill Lynch launched Project AMP, with additional DOE funding, to finance 750-megawatts of distributed generation. For America's large-scale on-site generators of solar electricity, *it's all about the money*. Today, the largest users of solar PV in America are, after Wal-Mart: Costco,

Kohl's (150 stores), Walgreens (200 stores, also planning 1,000 solar installations by 2020), Macy's, Staples, Bed Bath & Beyond, Toys "R" Us, GM, Ford, FedEx, Toyota, Whole Foods, Anheuser-Busch, Safeway, Lowes, PepsiCo, IBM, DuPont, Alcoa, Dow Jones, Johnson & Johnson, Intel, HP, Google, and eBay. IKEA has already put solar on 89 percent of its stores. Apple is building huge solar farms in North Carolina to help power its data centers; the company is committed to providing 100 percent renewable energy to clean its cloud. This is just a tiny sample of America's thousands of commercial solarelectric producers and consumers.

New Jersey, now with a gigawatt of installed solar, unexpectedly became the second state to lead in solar installations when it perfected the market for solar renewable energy credits (SRECs) as required by its ambitious renewable portfolio standard (RPS). An RPS requires utilities and the operators of the electric grid to either purchase solar power directly or buy SRECs, which finance all forms of solar to reach the state's goal of 22.5 percent solar by 2021. Twenty-nine states, plus D.C., have instituted RPS programs.

Arizona, Colorado, and New Mexico are ranked next, thanks to abundant sunshine, which generates more bang for the solar buck, thus attracting investment. Colorado's Proposition 37 created the state's RPS by popular vote when the legislature stonewalled it, and solar took off. Colorado recently launched its Solar Rewards Community, which enables customers of Xcel Energy to buy access to solar power without the need to install PV panels on their residences and businesses. Colorado expects to reach one million roofs by 2030.

The above states are followed by Nevada, Massachusetts, New York, Delaware, and Pennsylvania. Hawaii is next, where solar has reached grid parity. Maryland, with its booming solar market, didn't make USA Today's top-ten list of states with the most installed PV, but it did make the Greentech Media (GTM) report's list—these numbers fluctuate annually. Even little states are making a difference: Vermont has become the leader in streamlining the permitting process for small-scale solar. Community solar farms and "solar gardens," allowing multiple individuals to share in the

A 50-megawatt solar power plant using large trackers in Colorado's San Luis Valley, one of several massive solar farms near Alamosa. *(Author)*

benefits of a single solar installation, are popping up in Utah and elsewhere thanks to special tax policies, favorable regulations, and innovative financing. Abandoned industrial sites (brownfields) are being converted to "brightfields" at the local level. SolarMosaic is a young Bay Area company developing the community-solar-business model featuring "virtual net metering" and "crowd-sourced" funding.

Solar is proliferating on the rooftops of the nation's ballparks and stadiums, including the homes of the Boston Red Sox, the Kansas City Royals, the St. Louis Cardinals, and the Cleveland Indians. A huge solar array covering the parking lot supplies FedEx Field in Maryland, home of the Redskins, with 20 percent of its power.

In Louisiana, Brad Pitt's Make It Right Foundation built 150 homes in New Orleans, all with solar systems on their roofs. "New technology isn't just for the rich," Pitt says.

In Texas, the cities of Austin and San Antonio in 2012 both struck deals with investors and manufacturers to build 400-megawatt solar plants. Austin Energy opened its 35-megawatt PV power plant in 2011.

Believe it or not, more solar jobs were created in Connecticut and Illinois in 2012 than in any of the top-ten solar states, so go figure! And interestingly, "red states" lead in green jobs as a percentage of the workforce, and they also lead in green-tech job growth (there are now more than 2.5 million green jobs in the United States).

In Huntsville, Alabama, local solar advocates are working on creative solar solutions for the city. Redstone Energy is building solar-covered parking structures. "We believe the vision of a car being charged by sunshine and not buying any foreign oil is a great vision of our future," said Redstone's president. Indeed!

Unfortunately, "overreach" by federal regulators has halted the Property Assessed Clean Energy (PACE) program in twenty-five states. PACE enables the cost of energy efficiency retrofits and solar home systems to be paid out of annual property tax assessments, a sterling idea if there ever was one. Many states are going ahead with their own version of the program independent of the Federal Housing Finance Authority.

The loser states are southern: Florida, Virginia, Georgia, mainly. I won't go into the bizarre politics of solar in Virginia or Georgia, where Dominion Energy and the Southern Company hold sway. Things, hopefully, may have changed by the time you read this, for solar potential is huge in these states. Atlanta already boasts two large crystalline-PV-manufacturing companies—Mage Solar and Suniva—serving East Coast markets. As mentioned, Georgia has revised its rules regarding third-party solar generation.

It is the so-called Sunshine State that is the home of solar schizophrenia. Florida is solar's basket case. Florida Power & Light wants to build all the solar themselves (the company has installed 110 megawatts of PV). So they have lobbied against legislative initiatives that would encourage independent residential and distributed power generation, including the third-party financing that is behind the solar boom in twenty-nine states, creating tens of thousands of jobs. Not in Florida. The 2012 energy bill out of Tallahassee stripped out prior directives that were intended to create a state renewable portfolio standard (RPS) and a system of tradable and financeable solar renewable energy credits (SRECs). An extreme-right Tea Party governor didn't help. The respected Florida Solar Energy Center at Cocoa is forbidden to advocate or lobby for solar and is basically useless (unlike the North Carolina Solar Energy Center, which has influenced national and state policy through its outreach programs).

At the same time, it's not all bad news. FPL plans a 75-megawatt PV solar farm at the proposed all-solar city on Southwest Florida's Babcock Ranch development. The Florida utility brought its 75-megawatt Martin CSP plant online in 2010. Oddly enough, FPL has built more sun power plants than any other American utility; its parent, NextEra, is planning a 750-megawatt PV project near Palm Springs. In addition, NextEra is the largest owner of wind power in the country. FPL, to its credit, is the cleanest utility in the United States and charges the lowest kilowatt-hour rates in America.

Florida-based solar companies like Regenesis continue to build multimegawatt PV projects for municipal-owned utilities and universities, including Orlando's six-megawatt PV project to supply its city utility. Blue Chip Energy, based in Lake Mary, Florida, is presently completing a 100-megawatt solar farm. Blue Chip's subsidiary, Advanced Solar Photonics (ASP), manufactures solar panels proudly advertised as "Made in America" to supply its vertically integrated market, right down to its "SunHouse" residential line. Nearby, also based in the same Orlando suburb, is one of America's largest turnkey solar developers, the Spanish firm ESA Renewables, which is financing and constructing large-scale PV plants around the United States.

Meanwhile, a 400-megawatt project requiring an investment of $1.5 billion is planned for the Panhandle by National Solar Inc. in partnership with Progress Energy (now part of Duke Power). It would generate enough power for 32,000 homes. (I hope my skepticism toward these much-hyped PV megaprojects is misplaced.)

America's largest solar power trade show and exposition, sponsored by the Solar Energy Industries Association (SEIA) and the Solar Electric Power Association (SEPA) was held at Orlando's Orange County Convention Center in September 2012, with over 17,000 people in attendance and 950 exhibitors (the center itself has 1 megawatt of PV on its roof). I was excited to attend Solar Power International 2012 and was astounded at the compound growth of the global solar industry since I sold my interest in my company in 2008. Here were huge exhibits filling the vast halls by GE Energy, Sharp, Panasonic, Hyundai, Kyocera, Mitsubishi, Westinghouse, LG, Honda, Spain's Isofoton, and Abengoa, as well as the French nuke plant builder, Areva, now into solar. Here were all the Chinese firms like Suntech, Trina, Yingli, JA solar, Jinko Solar, Canadian Solar, Ningbo, Renesola, and a bunch I'd never heard of, their displays all staffed with knockout "booth babes" assumed to be more interesting to look at in their very short skirts than solar panels and hardware.

Here, too, were the American market leaders: SunEdison, SolarWorld, Bechtel, Outback Power, SMA America, Firestone, and Solar City, now with two thousand employees nationwide. I came away from the show (as I had the previous year in Dallas) convinced that solar is in our future more than most Americans can imagine, that it's coming faster than we know, and that everything in the cleantech, clean-energy, and sun-power business is vectoring upward so fast that writers and reporters cannot keep up or begin to understand the magnitude or velocity of the change. Just know it's coming, and it's all good.

President Bill Clinton addressed the packed auditorium of solar professionals, maybe ten thousand in all. At the Clinton Global Initiative three years earlier, I had extended personally, on behalf of SEIA, an invitation to Bill to speak at this annual event, but it took time for schedulers to work it out. We were all thrilled that he made it, showed up on time, and spoke for over an hour.

President Clinton outlined the facts about the "fastest-growing industry in the United States." He said, "Most people don't know that one hundred thousand people work in solar. In the second quarter of this year, solar power grew 116 percent over the same period last year."

"To listen to people on the other side of this debate," Clinton said, "you would think the president and his allies in Congress totally robbed the Treasury to subsidize bankrupt industries. But there are still twenty-two dollars in subsidies for coal, oil, and nuclear for every tax dollar invested in clean energy. These are things people need to know."

He mentioned Solyndra, of course, noting that it represented less than 1 percent of the money the Department of Energy has directed toward solar power. "People make mistakes," he said, observing that if we had shut down our space program every time a rocket or space shuttle blew up, we would have never gotten to the Moon or to Mars.

Rhone Resch, SEIA's CEO, in his opening remarks, highlighted the politicalization of renewable energy by the Republicans. "Our industry is under attack by the fossil industries generally and the Koch brothers specifically," he told the audience, adding, "but we are not an issue; we are an industry." He said that 80 percent of the negative conservative super-PAC attack ads running up to the 2012 presidential election had assaulted renewables. He pointed out that presidential candidate Mitt Romney and his running mate, Paul Ryan, had promised to roll back solar's 30 percent investment tax credit and end all incentives for renewable energy. We were all very worried.

Rhone asked Bill Clinton about the upcoming 2012 election's potential effect on the solar and wind industries, and Clinton answered, "Politicians are more honest and hardworking than you think, and they pretty much do what they say they are going to do. So listen to what they are saying." Thankfully, enough people listened.

McKinsey's Global Solar Initiative Department reports that it expects PV to transform power markets and says solar could potentially see investments of $800 billion to $1.2 trillion over the next decade. One technical study calculated that if all the qualified and available roof space in America—that is, 25 percent of residential and 60 percent of commercial—were covered in solar, its potential power production would reach 712 gigawatts, or two thirds of America's total generating capacity! We may just get there since solar is already on a rapid growth trajectory that can't be stopped. Add in strong energy efficiency and conservation programs that could realistically cut our electric power usage in half, and solar can do the rest.

Despite the reduction in federal support and solar subsidies, it is because of state leadership and intelligent, forward-thinking Americans in our towns and cities that independent, distributed generation and clean-energy entrepreneurship continue to explode, creating more jobs and benefiting the economy beyond anything I could have forecast a few years ago. When the people lead, politicians follow. By the time you read this, your own community may have reached the tipping point, when solar power is cheaper than the grid, and solar won't be a do-gooder energy choice anymore.

My reason for starting Standard Solar was the recognition that the largest untapped solar market in the world is U.S. residential rooftops. It still is. There are millions of roofs waiting to become energy producers. Solar City also figured this out early. Despite the increase in multimegawatt PV projects, *distributed generation,* in other words, home and commercial installations, leads utility-scale solar year after year by at least 30 percent. People want to generate their own solar power; it's as American as apple pie.

Thousands of licensed solar practitioners in America specializing in home and small commercial solar systems have seen the opportunity and are responding to the demand. They own or work for small "mom-and-pop" operations—mostly "pops," although there is a substantial number of women working as PV technicians. Most installers are credentialed by the North American Board of Certified Energy Practitioners (NABCEP), following a rigorous training and testing procedure. Many are also licensed electricians. The solarelectric retail and distribution business in North America today is in the hands of small "downstream" players, not the large corporations and manufacturers. These are the people Scheer calls "practical protagonists," the ones who do the work of solar, not just talk about it. These are America's new energy entrepreneurs, who will "democratize" our electricity system. Tom Friedman predicts the future is in energy entrepreneurship meeting energy technology, or "EE meets ET," perhaps a simplistic slogan, but he's right.

What America needs are more "solarpreneurs" and a thousand Standard Solars, not necessarily with as much capital or as many employees, but big enough to provide professional services in their local market. Roofers, electrical contractors, and HVAC firms are discovering the huge opportunities in solar electricity, and plumbers are learning that solar hot water can be profitable. If I were a young semiskilled worker, a high school or college grad, I can't think of anything better to do that would guarantee a lifelong income and possibly a big payoff than learning solar and becoming a dealer/installer—you don't need much capital to get started. (See Appendix 2 for more information.) *The money is in solar retail, not manufacturing, and the jobs are here, not in China.*

More solar education would help, especially at technical schools. Only two universities I know of—the University of Massachusetts, Lowell and Penn State University (online)—offer courses in solar engineering. In Germany, almost every university offers a master's degree in renewable energy. However, you don't need to know solar or electricity to start a solar business.

Witness Phil Coupe and his partner, Fortunat Mueller, in Maine, who had no training in solar, so they teamed up with an older, experienced practitioner, William Behrens, and founded ReVision Energy LLC ten years ago. Today, ReVision has forty-two employees, several branches in Maine and New Hampshire, a fleet of trucks, and Chevy Volts for sales site visits, powered by solar charging stations. The company weathered the 2008–09 recession, never laid off anyone, maintained solid growth, and is now the largest solar-PV-installation company in northern New England.

I asked Phil Coupe why he did it. "I have three kids, and I was worried about their energy future, about security, pollution, and geopolitical conflict. I wanted to help influence a better outcome." Maine is a poor state, and not a sunny one, but no matter. "There are plenty of committed people here who want to buy solar, and believe it or not, Maine has 33 percent more sunshine than Germany!" he told me.

Other entrepreneurial successes, out of thousands, are Phat Energy in Pasadena, California, and Roof Diagnostics in New Jersey, both with nutty names, but they are experiencing instant growth and success. The former was founded by my friend Phil Hartley, a French Moroccan film producer, who was inspired by my earlier book, he insists. I think he already had the bug, but he had no training or experience. Instead, he trained himself by putting solar on his own house, raised some angel funding, and figured out creative ways to finance residential systems. He then addressed a unique niche market: solar patio roofs and carports.

Roof Diagnostics, on the other hand, were experienced roofers who realized they were in a great position to install solar (let's face it: electricians should be jumping into the solar business, but many don't like to climb on roofs). "We turned a roofing company that did solar into a solar company that does roofing," says Kelly Pegler Jr., who has expanded to New York and Massachusetts. "We really are a solar company now."

I'm impressed and inspired every time I meet solar entrepreneurs, such as Dave Hollister in North Carolina, owner of Sundance Power Systems. Some years ago, I chased down Dave at his Appalachian cove in the North Carolina mountains. Goats were in the yard; chickens ran through his sales office, located in a barn; and mules peered into my open car window as I approached his all-solar-powered compound up a long dirt road and parked next to his brightly painted company vans. I had known him at Greenpeace eighteen years earlier, where he ran "direct action" training camps and where his wife, Sierra, had worked for me. With little money but a huge drive to be independent, do good, and make a decent living, he started his little company with a few friends. Today, he's moved down into Asheville, with branches in Raleigh and Greenville, South Carolina, and is one of the region's largest solar-installation companies. He never sacrificed his values as he grew.

He told me, "People are choosing a different path, becoming conscious of their true natures, and aligning with higher values that guide them into a more peaceful and sustainable future in harmony with the planet we call home. Through this journey, we run face-to-face into our desires and attachments, into 'our way of life.' We need to make choices that reflect our newfound identities and values."

He reminded me of what Hermann Scheer said in a speech, "The fight for renewables is a 'no' for fatalism and a 'yes' for an everlasting future and a spiritual hegemony."

These are the ethics of the solar business, but don't let such a spiritually directed outlook turn you off. Sun power technology itself is agnostic, it's here, it's practical, and its growth is now exponential not because it is based on ideology or ethics, good intentions or spiritual values, *but because it is a better business proposition than burning fossil fuels.*

The indisputable good news is that solar power in most people's lifetimes will be the world's largest and most profitable industry. In 2013, Royal Dutch Shell released a report that predicted photovoltaic panels will be the world's main power source by 2070.

We *will* run out of oil and coal one day in the future, but we will *never* run out of sunshine (at least not for a few billion years), so we will never run out of electricity. Whatever happens in the future, we will always be able to keep the lights on and our autos running and our communications systems going through distributed and centralized solar power networks. We won't do it because we want to do it; we will do it because we have to do it. After all, businesses, economies, and nations want to survive and somehow always manage to.

Human consciousness is collectively at a tipping point. In a very short period China has leapfrogged from smokestack industries to clean tech. They know they have surpassed the United States in production of CO_2 emissions and they intend to do something about it. India's leadership is seriously concerned about doing something, too—not because of Kyoto or Copenhagen or Rio + 20—but because they know they must for their own survival. Europe is doing the most to reduce CO_2 because the public demands it.

In the United States, the evolution of the collective consciousness is greater than we know: it's organic, life-changing, bottom-up, inescapable, and is impacting us in ways we can't imagine. The politicians be damned; the people will lead and the government will follow. If in a few short years we can ban cigarette smoking nationwide, we can start decarbonizing the atmosphere. We can begin to reduce population growth. We may have hell to pay first, and may enter a mini-dark age along the way, but change *will* come. It always has, which is why our otherwise ignorant, murdering, selfish race is still around.

Each of us can choose our energy future, and we may have to as events force us to make a transition to a new energy paradigm and embrace the solar imperative. We can preserve a civilized way of life for ourselves and our communities by preparing now for a "postcarbon" lifestyle without sacrificing our way of life. *Economics, not environmental concerns, is now driving the uptake in clean, safe solar electricity.*

We can respond to the triple threats of global warming, energy security, and oil depletion individually or collectively. We can share or be selfish, act ethically or greedily—the sun doesn't care; it's there for all of us. As long as the sun shines and photosynthesis grows our crops and the photovoltaic effect and the wind make our power, what else do we need? When you've got sun power, every day is a sunny day.

And best of all, solar energy is *your* power. The sun is the engine of the earth and the only truly safe power source for the planet. The fires of our nearest star are there to work for us. The future is here, now, and it's brighter than ever.

APPENDIX 1

You've gotten this far without a technical discussion about how solar electricity actually works. Well, I don't know how it "works," but I can tell you how it functions and performs—what PV does, not how it does it.

First, it is useful to understand that sunlight hits the earth at a "constant" of 1,350 watts per square meter, and it makes its way through the atmosphere to strike the earth when it is directly overhead ("peak sun") with the energy of 1,000 watts per square meter. Photovoltaics can capture only a fraction of this, depending on the efficiency of the silicon solar cells used. A meter-square solar module, not counting the wasted space between the cells, could effectively capture 150 watts at "peak sun" if it was 15 percent efficient. Most modern crystal-silicon solar cells are between 15 percent and 22 percent efficient. In the laboratory, some cells have proved to be as much as 30 percent efficient, but production cells have yet to achieve this. Nonetheless, capturing 15 to 20 percent of all this free energy is no small accomplishment.

The amount of power a solar module, commonly called a "panel," puts out is calculated in "watts peak," or Wp (the abbreviation of watt, W, is usually capitalized in honor of Scottish mathematician and engineer James Watt). A 35-Wp module will reach that power output in peak sun.

The next thing you may not need to know, but which I'm going to tell you, is that, curiously, a solar cell of any size or type makes half a volt of direct current (DC). Don't ask why; that's just the way it is. Although solar modules are rated in watts, which relate to amperes, or amps, the important thing, at least in the off-grid business, is to know that the panel, whatever its size, has enough volts to charge a battery. Most large batteries work on 12 volts. Another fact: it takes roughly 17 volts to charge a 12-volt battery. Thus, a solar module must have 36 cells (36×half a volt, or 0.5) to put out approximately 18 volts. None of this is precise because battery science is a black art, based on arcane electrochemical principles that have been studied for over 150 years. No two battery experts agree on how best to charge a battery, so for those working in remote off-grid solar markets, it is always a matter of trial and error.

Batteries, especially the old reliable lead-acid type, are based on a 150-year-old technology. A good workhorse, "deep-cycle" (deep-discharge) battery can last seven to ten years. Nickel-metal-hydride batteries, developed and patented in the 1980s by the late Stan Ovshinsky's Ovonics Corporation, became the standard for golf carts, hybrid cars, and remote power systems, but they were too expensive for off-grid lighting systems in the developing world. Lithium-ion batteries and other technologies would break new ground in the late 1990s, but they remain expensive and even dangerous. Solar electrification would stick with lead-acid variations (sealed, gel cell, tubular, etc.) for years to come.

The other critical electronic part of any off-grid solar installation, big or small, is the charge controller. The controller's job is to make sure the battery does not get overcharged (too much sun) and does not get drained by using too much current. The controller will cut off the system when the battery drops below about 11 volts. Just as in a car, when a battery is drained below 11 volts it stops working. Solar cells produce only direct current, which is handy since batteries can only store direct current.

Special incandescent bulbs, radios, and TVs with DC inputs can run on direct current, but DC incandescents are impractical since they use too much energy. So, early on, the developers of off-grid solar lighting systems switched to fluorescent tubes, large and small, which are very efficient but run only on AC—a functional requirement of all fluorescent lights. So, small *inverters* were designed, which originally used 10 percent of the available power just to convert it from DC to AC. But the technology, based on demand, quickly improved. Small inverters were designed, coupled with a *ballast*, to turn on and run fluorescent lights. This small electronic device—the "inverter ballast"—was the critical part of solar lighting systems for twenty years. These, and the lights themselves, were always the troublesome part of a solar home system (SHS), since electronic inverter ballasts are prone to failure and early demise.

However, *sun power works*; what fail are the *parts*, known also as "balance of systems," or BOS. Solar modules always do what they are supposed to do, and if properly "encapsulated"—that is, if their cells are sealed between their metal "substrate" and glass—they will last at least twenty-five years and perhaps much longer.

When the compact fluorescent bulb (CFL) came along in the late 1990s, special ones were designed by Philips and Osram with inverter ballasts so they, too, could be powered by direct current and used in solar home systems. These were brighter than the old tubes, more reliable, lasted much, much longer, and used even less energy.

Then, the real lighting revolution happened: LEDs! Light emitting diodes have been around for a very long time, but only at the end of the twentieth century had they gotten bright enough, thanks to innovative manufacturers around the world, to act as a light source and even replace some conventional lighting. Not only have CFLs now become commonplace, in both DC and AC forms, but you can now buy LED lights at Lowe's, Home Depot, and Wal-Mart. IKEA states they will *only* sell LEDs by 2016.

LEDs use so little power and last so long that they soon revolutionized solar lighting in developing countries. And they work directly on DC current! They were, at first, very expensive, but even then it was often cheaper to use them because the same lumens could be produced with much smaller solar modules and batteries. Today, most SHSs sold in the two-thirds world come with at least two LED lights for focused lighting, complimenting the warmer CFLs that throw a wider room-area illumination. They're now made in every industrial country, and prices continue to fall; the LEDs themselves get brighter and brighter!

In the United States and Europe, LEDs are now beginning to be used for household illumination lighting. Meanwhile, it is the standard, rapidly evolving curly CFL that has now officially replaced most incandescent lightbulbs, producing huge energy savings for homeowners and businesses. CFLs constitute the biggest energy conservation breakthrough of our time (Republican lawmakers repeatedly try to kill laws mandating their use). Of course, these all work on AC, as do all of our appliances in the United States, Europe, and in cities and towns worldwide. So how do appliances that mainly work on alternating current (AC) work if solar electricity only comes in direct current form?

Again, *inverters*. But much bigger, more expensive, and now very efficient inverters; this was the breakthrough that created the *on-grid* or *grid-tied* market. Mostly made in Germany, Austria, Switzerland, and Korea, and now the U.S., inverter technology improved dramatically by the turn of the last century. Inverters' efficiency went from wasting 10 to 15 percent of the power it was "inverting" to less that 5 percent. This technical improvement—inverter efficiency—launched the European and American solar revolution. Manufacturers make inverters in all sizes, from 4-kilowatt for residential use and 500-kilowatt for solar farms. Inverter technology is really what is behind the high-growth solar energy business that doubles every year.

Much of this story is about small solar home power systems averaging 40 or 50 watts, but we have also talked about much larger PV installations from 6,000-watt or 6-kilowatt (kW) residential systems to 500-kilowatt commercial systems to 100-plus megawatt (MW) utility power projects. We talked about gigawatts (GW), which is a thousand megawatts. Then there are terawatts: a terawatt is a thousand gigawatts. Solar panels themselves are rated in "watt peak" (Wp), the power they will put out in full sun. A one megawatt peak (MWp) solar plant will supply power for approximately two hundred American homes, or twenty thousand homes in the developing world.

What else do you need to know? It's useful to understand kilowatt hours (kWh) and how they relate to prices, since conventional electricity is based on the cost per kilowatt hour, but solar systems are not (the apples and oranges quandary). At home you probably pay between nine and thirteen cents per kilowatt hour, unless you live in the Pacific Northwest, where hydropower still goes for seven cents per kilowatt hour. And you probably use close to 1,000 kilowatt hours per month, multiplied by the kilowatt hour price, which gives you your utility bill. However, if you have a business in California, you might be paying thirty-eight cents per kilowatt hour at "peak power rates," that is, midday, when industry, businesses, and air-conditioning are operating. Today, kilowatt hour costs for many American solar home installations work out to less than eighteen cents a kilowatt hour, which may seem higher than what their utility charges, but remember: by providing *free* power when the sun is shining, the system can pay for itself in as little as four to seven years with incentive pricing. Once paid for, the system will continue to provide free daytime power for another twenty years! And all that time, it is certain that your utility's electric rates will continue to rise.

Today, utility-scale PV power plants in the multimegawatt size are producing power selling wholesale for as little as six to seven cents (U.S.) a kilowatt hour, with tax incentives and depreciation. This means that solar power has nearly reached "grid parity"—when solar electricity's wholesale cost equals the retail price of conventional electricity—in states with high electric rates, which is the truly wonderful news that we have all been waiting for.

Sun power is now cheaper than conventional power if you have a way to finance the front-end costs of buying and installing it. The utilities build power plants with low-cost, thirty-year loans and then charge their customers a flat rate. Until recently, few households or businesses could get, or wanted, a thirty-year loan to purchase a solarelectric system. Solving this problem on behalf of customers in the developing world was the biggest challenge faced by companies selling solar, and it consumed the better part of eighteen years of my life. We all had to learn retail banking to sell solar to poor people in the two-thirds world. And what we learned in these countries about financing home solar I later brought back home to the United States, when I launched Standard Solar.

Meanwhile, how do you calculate kilowatt-hour costs with solar, you might ask? It's quite easy; even I can do it. Before I tell you, however, there is one more thing, which is actually the first thing: *insolation*. This refers to the intensity of the solar radiation striking the earth—that is, *watts per meter squared*—at any given place and time of day. Years of research have succeeded in mapping out the solar radiation, or insolation, all over the world. Typically, if cloud cover is not a problem, a solar module in the equatorial regions of the planet will receive six or seven "peak sun hours" a day. During those hours the sun is hitting that area with 1,000 watts of energy per square meter.

In the Northern Hemisphere, peak sun hours average three and a half to four and a half a day. In South Africa, Saudi Arabia, and western Australia, the sunniest places on earth, peak sun hours can reach eight, since there are few clouds. Weather and cloud cover are considered when calculating peak sun hours. Some of the cloudiest places on earth are right smack on the equator and receive only three peak sun hours per day. "Sunny Florida," with so many cloudy months, has fewer peak sun hours than Colorado, for example.

Now that we know the annual average peak sun hours available, we can multiply that number by the size of our solar module to determine how much power a solar home system in the two-thirds world will deliver. A 50-Wp module in a sunny place with five peak sun hours will produce 250 watt hours each day. Thus, six 9-watt lights will use 54 watts per hour (6 × 9), or 54 watt hours (a 9-watt compact fluorescent bulb, or CFL, provides the lumens, or light, equivalent to a 60-watt AC bulb, just like the compact fluorescent bulbs you buy for your home—says so right on the box). If you divide the available 250 watt hours, produced by the solar module and stored in the battery each day, by 54, you will see that the six lights can operate for 4.6 hours per night. A black-and-white TV connected to the system will use power equivalent to two lights (but measured in amperes), which is why people with battery-powered solarelectric systems turn off their lights when watching TV. A color TV uses twice the power of a black-and-white set. More recently, as previously described, LED light fixtures turn this formula on its head, since LEDs need only a fraction of the power required by a CFL.

We can now compare kilowatt-hour prices for the unelectrified and defeat the shibboleth that "solar is too expensive," the silly argument we fought against for twenty years (power is priceless when you don't have any). The 50-Wp system described above produces 250 watt hours a day, or 7,500 watt hours per month (30×250), or 7.5 kilowatt hours per month. The average American or European home, using approximately 1,000 kilowatt hours a month, consumes as much as 130 times more electricity than a solar-powered home in the two-thirds world. The ramifications of this are something to think about.

So, what is the cost of the 7.5 kilowatt hours produced by solar PV for the little house in India or Sri Lanka? That depends. If you amortize the $400 SHS over three years, or thirty-six months, plus 10 percent interest (a normal rate in these countries) that's $550, divided by 36, or $15.00 per month, which works out to $2.00 per kilowatt hour. Development economists claimed for years that this was too expensive, comparing it to the cost of grid power, which wasn't available! Economists, practitioners of the "dismal science," are people who lie awake at night trying to figure out if what works in the real world might actually work in theory. In the real world, the 7.5 kilowatt hours of DC running DC appliances and lights provides almost the same comfort and convenience to a poor family in the developing that a first world family enjoys, minus the refrigerator and HDTVs. This was even possible when solar panels were selling for $3.50 to $4.50 a watt, which were the prices we paid in the 1990s. At today's prices, these same poor families, unserved by the grid, will soon be able to power wide-screen TVs and energy-efficient refrigerators. Many are already using computers.

If we look at "life cycle" costs (which economists like to dwell on) that give the SHS a fifteen-year average life (less for batteries and electronic components, more for the PV module) and divide 180 months into $550, then monthly costs are $3.00 for 7.5 kilowatt hours, or forty-eight cents per kilowatt hour. But forget kilowatt hours!

These people aren't using and don't need kilowatt hours—they are using *watt hours,* efficiently and productively. And they get all the watt hours they need for $3.00 a month, amortizing the system price plus interest over 180 months. In the real world, people are quite happy to light up their world for $3.00 a month.

Even the poor can afford this, especially if each member of a family of six earns one dollar a day, which means they bring home $144 a month (working six days a week). They don't pay taxes because they are too poor, and because they are too poor the government or private utilities won't hook them up to the power grid should it come along, since hookups are expensive—usually a third of the cost of an SHS—and it's not worth it to the utility unless the family uses at least 300 kilowatt hours a month, which, even at a subsidized five cents a kilowatt hour, will cost them fifteen dollars a month. So which is cheaper, solar or conventional fossil-fueled grid-delivered power? Never mind the hidden subsidy in all conventional power delivery; if the family had to fund the amortization of the government's power plant over three years, the kilowatt-hour price would be more like five dollars, not five cents. Instead, the government probably got the power plant for free, thanks to a multilateral donor or international investment group, or funded it with tax money on thirty-year terms.

In chapters 4 and 11, I have already discussed how to compute residential-solar-system costs and power production for a grid-tied American home.

There are a couple more technical items to discuss: production technologies and energy storage. The three main types of PV are monocrystalline, polycrystalline, and thin film. I'll keep it simple. Monocrystalline, or "single crystal," requires solar-grade silicon to be "grown" or extruded into long ingots in hot furnaces. These ingots are then sliced into wafers from three to five inches square. This is how all computer chips are made, in case you wondered. It's the same stuff.

Polycrystalline, the most common technology, uses waste silicon from the computer industry with imperfect crystal formations. It is melted into ingots, which are sliced into paper-thin wafers by hair-thin wire saws. The wafers, either poly or mono, are then "doped" (chemically treated) to function as photovoltaic cells, which convert light hitting their surface into electrons that can be drained off by wire "tabs." These cells are strung together and mounted on a plastic-coated steel "substrate"; then "tabbed" together in series and covered with a piece of low-iron tempered glass; and finally "laminated" under high heat to bond with the plastic layer (DuPont Tedlar) beneath the cells. A metal frame is attached, and a solar module is born. Most of this process is automated and often robotized. Making solar cells is an exquisitely refined process and is fascinating to watch.

Thin-film technology is also called "amorphous," and there are several types, the latest being copper-indium-gallium-selenide, or "CIGS," and cadmium-telluride (Cad-Tel, or CdTe). These contain much less silicon, reducing costs, but produce only half the efficiency of "poly" or "mono" PV panels. Thin-film manufacturers were the first to bring per-watt costs down to one dollar per watt, but now poly and mono producers have nearly matched that number. Ken Zweibel is one of the unsung heroes of thin film, developing cost-saving and materials-saving processes at the National Renewable Energy Laboratory thirty years ago, which developed into a spin-off company, Primestar, that licensed the DOE-sponsored research. GE bought Primestar in 2007, renamed it GE Solar, and aimed to get the technology to 14 percent efficiency. However, GE dropped its plans for a 600-megawatt cad-tel manufacturing facility in Colorado when it saw the writing on the wall: polysilicon had become the least-cost PV technology. GE switched to building solar power plants with traditional polysilicon PV panels.

Zweibel moved to D.C. to launch the Solar Energy Institute at The George Washington University. He predicted in a 2008 *Scientific American* cover story that solar could provide 69 percent of America's electricity by 2050, displacing fossil fuels and nuclear power. I believe we'll get there even without thin films.

Thin film, both cad-tel and CIGS, are a problematic process, and many large production facilities favoring this approach have subsequently been shut down, the most recent being United Solar Systems

(UniSolar), founded by the great pioneer Stan Ovshinsky. The other major thin-film player, First Solar, for a period the world's largest solar company, saw its stock drop 88 percent in 2012, although it somehow still thrives; it has deep pockets. The idea originally was that, instead of the time- and labor-consuming process of making silicon wafers, and to reduce the silicon content, it is possible to deposit small amounts of silicon and all the other chemicals and bonding agents in a gaseous form directly on metal or glass. Since this can be entirely automated (although using extremely complex and expensive machinery), and since glass is cheap, it was thought this would be the way the world would be covered with solar. But solar-grade silicon prices fell dramatically as the silicon-refining industry ramped up quickly in 2005 and 2006.

At the same time, since thin films are only half as efficient as crystal silicon, it takes twice as much component glass and aluminum, and more land and bigger roof space, to install the same amount of power. And despite the many millions spent by pioneering companies like UniSolar, France's Photowatt, and Shell Solar, the per-watt price equation never changed. Meanwhile, CIGS thin films are poised to replace cad-telluride as higher efficiencies are being developed. The Chinese company TCMS Solar has bet its future on CIGS technology. But it's still not clear if thin films will be around in the future.

Notwithstanding the problems of thin-film solar, the infamous Solyndra in California tried a new form of thin-film technology, which almost everyone in the industry knew was a bad idea, but gullible investors, and the naive Obama administration (and the fawning press) were fooled, and you read the rest of the story in the newspapers in 2011 ad nauseam and saw TV commentators beat the story to death. The Republicans used the $500 million loan guarantee lost in the Solyndra debacle to slam the Democrats, somewhat deservedly. Unfortunately, they also used the Solyndra debacle to attack the government's green-energy programs, most of which have been hugely successful. The other thin-film loser, a darling of naive investors and journalists, was a company called Nanosolar.

Thus, the workhorse of PV has become the good old monocrystalline and polycrystalline silicon solar cell, and no "new" photovoltaic technology has come close to replacing it. Many have tried and more are trying, but I'd rather believe there really is nothing new under the sun when it comes to basic PV.

The polycrystalline cells are especially appealing to the eye because they are cobalt blue and sparkle in the sun like magic jewels. "Single," or monocrystalline, cells are generally black, not so pretty, but can be more efficient: in 2007 the manufacturer, Sunpower Corporation, thanks to the tireless lab work of solar research giant Richard Swanson, created the world's most powerful PV panel in commercial production. He developed a single crystal cell rated at 22 percent efficiency (the latest are 24 percent efficient). But others, like Suntech, Panasonic, and Sharp, have recently caught up.

I should point out that it takes power to make PV, of course. This is known as EROI—energy returned on energy invested. Nothing is free, not even free power from the sun! It is commonly accepted that it takes the power equivalent of one year's output of one PV module to produce that same module. So from a module guaranteed by the manufacturer for twenty-five years, you really only get twenty-four years of free electricity. Not bad. Using the most efficiently produced solar cells with thirty-year expected lifetimes, the EROI works out as high as 60 to 1.

Karl Wolfgang Böer, after a lifetime of groundbreaking solar research at the University of Delaware (as discussed in chapter 12), told me that he believed that PV modules meeting the highest technical standard of "encapsulation" may last one hundred years. In addition, Ken Zweibel points out that solar panels can be recycled. A PV expert recently reported that the output of a 33-Wp ARCO solar panel had only degraded 8 percent since it was manufactured thirty years ago. We make them much better today.

The other "big" solar technologies are concentrated PV (CPV), where sunlight is focused by lenses on photovoltaic cells, increasing output (see chapter 12), and concentrated solar power (CSP), a solar energy technology unrelated to PV. There are two types of CSP. One technology developed long ago focuses sunlight from tracking mirrors, called heliostats, on a receiver atop a huge "power tower," where the heat makes steam to run turbines. The more common CSP technology, invented by Frank Shuman, who built the first one in Egypt in 1910, features polished parabolic troughs that focus sunlight on tubes filled with heat transfer fluids that make steam for conventional generators. These also require tracking systems, adding more expense. Both kinds of CSP power plants can be tied easily to thermal energy storage (TES—read on), which is the only advantage of keeping CSP alive these days. CSP comes in one size: gigantic. This appeals to utilities that like to think big and can't get their minds around "distributed generation," which is the hallmark of photovoltaics, a technology that works at any scale, centralized or decentralized, from 40-Wp solar home systems in the developing world to 100- and even 1,000-megawatt power plants in the American desert in Nevada and Arizona (the Tonopah and Ivanpah projects). Despite the problems and costs inherent with CSP, the French nuclear reactor builder Areva launched Areva Solar and Areva Wind in 2011 and built a 250-megawatt concentrated solar power (CSP) installation in India's Rajasthan state.

While much of this story is about how people with no electricity around the world are getting their power from the sun, the book is also about how we can, too, and how the solar revolution is under way in the industrial world. This enormous global grid-tied market inherently involves utilities, some of which imagine the future and cooperate with solar advocates and promoters, and some who fight it. In this regard, the issue of energy storage and "base-load power" necessarily comes up. *Energy storage is the Achilles' heel of solar* (and is also the greatest business opportunity).

The problem with solar and wind, of course, is "intermittency." Solar only operates at 35 percent capacity—that is, the daylight hours. The variable output of power from residential and commercial "distributed generation" poses a problem of "uncertainty" to the electric utilities. Gerry Braun, my first partner at Standard Solar, says, "I prefer '*variable*' to intermittent. I think 'intermittent' could well be a marketing term like 'alternative energy,' intentionally designed to make you not want the product."

Electricity supply and demand must always be in balance in real time. "What is changing with rooftop solar is that we haven't dealt with lots of solar generators in a serious way," says Silicon Valley investor Vinod Khosla. Some people think the so-called smart grid will help solve the problem. Khosla thinks not: "It's more smart hype than smart grid," he says. "Smart meters do not mean a smart grid. We need intelligent power electronics, a huge new opportunity." He noted Moore's law, according to which microchip performance doubles every eighteen months, and compared it with "Westinghouse's law," where, he said, "big changes in the electric grid occur every fifty to seventy years."

Khosla believes energy storage is the better bet. Utilities want to manage uncertainty. "Storage is the key element of certainty," he told the Electricity Storage Conference in 2011. "Your task and role is far more important than people realize." He noted that batteries are not the answer, except for off-grid homes in the developing world. Sitting next to John and Annie Glenn at a Clinton Global Initiative renewable energy workshop, I asked America's greatest astronaut what he was doing now. "I'm at Ohio State working on energy storage." He was close to ninety and focused on clean energy solutions!

Knowing nothing about utility-scale energy-storage technologies, I met with Joe Simmons again, the aforementioned director of the Arizona Research Institute for Solar Energy. He previously advised Gabrielle Giffords on solar energy policy in Arizona, and she actively pushed in Congress for the 30 percent federal tax credit on solar PV, which remains through 2016. Joe, who is replicating the Arizona solar institute at Florida's Gulf Coast University, told me about compressed-air energy storage (CAES) and gave me his exhaustive study "CAES with Grid and Photovoltaic Energy Generation," which he produced for the Arizona Research Institute. It's not new. It's been used for decades in Germany and Alabama. The Alabama plant built by Dresser-Rand in 1991 uses electric power to compress air into dry underground caverns. When the air is released it drives electric turbines, producing 110 megawatts of power for up to twenty-six hours. Limestone mines, aquifers, or even large storage tanks also work. The largest in the United States, which will store 2,700 megawatts, is being built in Ohio. The DOE's Sandia National Laboratories are developing CAES technology to work directly with wind and solar. CAES can come online in fourteen minutes, matching the daytime input from the variable renewable sources. Want to make money? Do some research on companies or utilities building CAES plants to stabilize solar and wind generation, and invest now! Vinod Khosla is; he didn't become a billionaire betting on the wrong things. His latest play, with coinvestor Bill Gates, is LightSail, a CAES venture that seeks to revolutionize energy storage. Google is also getting into the energy storage game.

However, the caveat is that many experts believe the utilities can absorb up to 30 percent "solar penetration" from grid-tied solar before they need to worry about intermittency and storage. "Utilities have no imagination," says one PV project developer, who believes that demand response, remote monitoring, efficiency, and other "smart grid" solutions can be brought to bear on the issue of variable solar versus base-load power. Our future energy mix from renewables will clearly be a mix-and-match proposition offering huge opportunities for innovation and the prospect of great fortunes to be made.

The oldest energy storage system is hydro. Excess solar and wind power is used to pump water uphill to storage lakes or tanks in the daytime, to be released at night. But this can't work everywhere. The TVA has been doing this in Tennessee for over half a century. The other big storage technology is thermal energy storage (TES), used by utility-scale concentrated-solar-power systems (CSP). TES can store molten salt underground, releasing it to produce steam for running turbines at night. PV is now beating CSP on construction cost, but PV doesn't offer the easy energy-storage potential of CSP. Also, CSP has environmental issues, and many moving parts. Developers of a 1,000-megawatt CSP project in California switched to PV in 2012 to save on costs, but they will have to look to other forms of storage.

However, some analysts see a bright solar future in which CSP and PV will complement each other, utilities using PV to produce their cheapest power by day and relying on CSP to provide their day-and-night base-load power, thanks to TES, which is more efficient and cheaper than batteries or CAES. Are you confused yet? So is everyone in the solar industry and its financial backers, many of whom don't know which horse to back. I pointedly didn't talk about flywheels here, another hopeful storage technology.

Maybe there is nothing new under the sun, but there are many competing solar energy technologies, and the most cost-effective one will dominate. Frankly, I think it will be a combination of decentralized, distributed residential and commercial PV, huge centralized solar farms, and some utility-owned and -operated CSP plants. The 1–100-megawatt solar farms are coming online worldwide faster than anyone can count or that I can list here, while CSP is still facing locational and financial hurdles.

Despite Khosla's and others' reservations about batteries, AES Energy Storage is building a 32-megawatt lithium-ion-battery storage system at its wind farm in West Virginia. Solar City and Tesla Motors are installing lithium-ion residential storage systems, and Panasonic and Hitachi are selling home-based, solar-backed energy storage in pilot projects. Duke is building the largest battery storage facility at its Texas wind farm. Anticipating the business opportunity of the "smart grid," Samsung has teamed up with Bosch to market battery-based residential and community modular energy storage systems.

A brand new venture, Solar Grid Storage, launched by longtime solar leaders Tom Leyden and Chris Cook, allows modular lithium-ion-battery banks to be integrated with solar PV installations through an innovative financing model. The company adds energy storage to grid-tied PV—up to 10 megawatts—that will enhance grid stability and provide standby power while alleviating the "intermittency" inherent in distributed generation, a big concern of grid operators like PJM Interconnection in the Northeast.

Peak power is now cheaper from solar and wind—although gas may give them a run for their money—so why not save some of that power to feed back to the grid at night? Extreme Power in Texas is building megawatt-size storage batteries with a mysterious proprietary technology coupled with power electronics that can store and deliver power for $1,000 per kilowatt, equal to the cost of turning on a peaking power plant to serve daytime peak demand. Extreme Power, which was formed originally to serve the electric-car market, also sees their batteries in electric vehicles, including fleets and trucks, as storage devices for homes and industries, providing the intelligent power electronics and communications protocols that can be added to the energy management mix. This is not a new idea (Amory Lovins envisioned it twenty years ago), but it still remains in that pie-in-the-sky category. We need millions of electric cars first, and clean power to charge them.

Perhaps the most exciting "breakthrough" that might actually work is liquid-metal-battery storage, invented by solid state chemist Donald Sadoway in league with his students at MIT. *Time* magazine recognized Sadoway as one of America's 100 "most influential people in the world" in 2012. Says Sadoway, "The battery is the key enabling device so we can draw electricity from the sun even when it's not shining." Focusing on the intermittency problem with renewables, he believes we will "invent our way out of the energy crisis." He says, "think big and think cheap," adding, "if you want to make something dirt cheap, make it out of dirt." His electrometallurgical materials are as earth abundant as PV's silicon. He believes his batteries, to be produced by his new corporation named Ambri and funded by Bill Gates, Khosla Ventures, and France's Total, are the grid-level storage solution everyone has been waiting for. One battery fitting into a forty-foot container will store 2 megawatts of electricity, enough to supply two hundred homes. They are, he says, priced to sell without subsidies, emissions-free, and remotely controlled and have no moving parts. Just like PV! This just may be the answer. To learn how they work—or perhaps get a job—google Donald Sadoway.

Thinking about the so-called smart grid that can integrate distributed solar and wind with conventionally generated electricity and energy storage got me recalling what Tom Tatum said to me back at the DOE in 1980: "In the future, every home will have a microprocessor to manage its power supply and demand." Westinghouse's law still gets in the way, but we *are* making progress, and distributed sun and windpower *are* the future. Technology is not the problem. Political will and public resolve are what is needed, which is lacking in the United States but commonplace in Europe and China.

In any case, the silicon solar cell has become to the twenty-first century what the silicon microchip was to the twentieth. Semiconductor materials may be the new gold. Fortunately, silicon is globally abundant, and so is sunshine.

APPENDIX 2

Additional Sources of Information

Websites and organizations including a few select private solar installation, sales and service companies:

Data Base of State Incentives for Renewable Energy
www.dsireusa.org

American Solar Energy Society
www.ases.org

Solar Today magazine (published by ASES)
www.solartoday.com

Solar Energy Industries Association
www.seia.org

Solar Energy International
www.solarenergy.org

National Renewable Energy Laboratory
www.nrel.gov

International Solar Energy Society
www.ises.org

North American Board of Certified Energy Practitioners
www.nabcep.org

PV Magazine
www.pv-magazine.com

Home Power magazine
www.homepower.com

Interstate Renewable Energy Council
www.irecusa.org

Solar Decathlon
www.solardecathlon.org

Greentech Media
www.Greentechmedia.com

Solar Buzz-Solar Market Research & Analysis
www.solarbuzz.com

Renewable Energy World
www.renewableenergyworld.com

Pixy Jack Press
www.pixyjackpress.com

Standard Solar Inc.
www.standardsolar.com

Revision Energy Company
www.revisionenergy.com

Phat Energy Company
www.phatenergy.com

Solar City Inc.
www.solarcity.com

SunSpot Solar Energy Systems
www.sunspotenergy.com

Sundance Power Company
www.sundancepower.com

Energy Efficiency & Renewable Energy, U.S. DOE
www.eere.energy.gov

U.S. Energy Information Administration
www.eia.gov

OnGrid Solar; Solar Electric Financial Analysis
www.ongridsolar.com

Solar Business Alliance
www.solarbusinessalliance.com

Blue Green Alliance
www.bluegreenalliance.org

Planet Forward: New Ideas in Sustainability
www.planetforward.org

Jeremy Leggett's blog
www.jeremyleggett.net

Solar Century, UK
www.solarcentury.co.uk

Operation Free: Free America with Clean Energy
www.operationfree.net

Ontility: solar resources and training
www.ontility.com

Washington, DC Solar Tour
www.solartour.org

U.S. National Solar Tour
www.nationalsolartour.org

E2: Environmental Entrepreneurs
www.e2.org

Sun Wind Energy, Germany
www.sunwindenergy.com

Solar Electric Light Fund
www.self.org

Solar Electric Light Company India
www.selco-india.com

Carbon Nation, the movie
www.carbonnationmovie.com

Real Goods Solar
www.realgoodssolar.com

Whole Earth Catalog
www.wholeearth.com

Backwoods Solar Company
www.backwoodssolar.com

Vote Solar
www.votesolar.org

Solar Industry magazine
www.solarindustrymag.com

United Nations' Solar Energy for All
www.sustainableenergyforall.org

Renewable Energy Focus magazine
www.renewableenergyfocus.com

Mosaic Company: Grow Your Money with Solar
www.solarmosaic.com

Solgenix organization
www.solgenix.net

Rural Energy Delivery Companies Alliance
www.redcoalliance.org

Energy Access Foundation
www.energyaccessfoundation.org

Inter Solar Website
www.solarserver.com

Florida Solar Energy Center
www.fsec.ucf.edu

North Carolina Solar Energy Center
www.ncsc.ncsu.edu

Florida Solar Energy Industries Association
www.flaseia.org

Solar Television videos
www.solarpv.tv

Global Solar Alliance
www.global-solar-alliance.com

We Care Solar
www.wecaresolar.org

Solar Works for America
www.solarworksforamerica.org

Energy Fact Check
www.energyfactcheck.org

The Solar Foundation
www.thesolarfoundation.org

New Vision Renewable Energy
www.NVRE.org

Solar Town: Learn, Buy, Build
www.solartown.com

Solar Freedom Now
www.solarfreedomnow.org

Solar Tech: Silicon Valley Leadership Group
www.solartech.org

Institute for Local Self-Reliance
www.ilsr.org

Freeing the Grid: state net metering policies
www.freeingthegrid.org

Climate Daily: Web News
www.climatedaily.com

Community Power Network
www.communitypowernetwork.com

Clean Energy Authority
www.cleanenergyauthority.com

GLOSSARY

ACORE	American Council on Renewable Energy
AEP	American Electric Power Corp.
ASES	American Solar Energy Society
CAES	Compressed Air Energy Storage
CdTe	cadmium-telluride
CIGS	copper-indium-galium-selenide
CPV	concentrated solar photovoltaics
CSP	concentrated solar power (solar thermal)
DOE	Department of Energy
EDF	Électrcité de France
EPIA	European Photovoltaic Industry Association
EV	electric vehicle
EVN	Electricity of Vietnam
FPL	Florida Power & Light
GEF	Global Environment Facility
GNERI	Gansu Natural Energy Research Institute
GW	gigawatt
IPO	initial public offering
KV	kilovolt
kW	kilowatt
kWh	kilowatt-hour
LADWP	Los Angeles Department of Water and Power
NABCEP	North American Board of Certified Energy Practitioners
NGO	nongovernmental organization
NREL	National Renewable Energy Laboratory
NRG	NRG Energy
NYPA	New York Power Authority

OPEC	Organization of Petroleum Exporting Countries
OPIC	Overseas Private Investment Corporation
PEPCO	Potomac Electric Power Company
PG&E	Pacific Gas & Electric Company
PV	photovoltaics
PVMTI	Photovoltaic Market Transformation Initiative
RBF	Rockefeller Brothers Fund
RPS	Renewable Portfolio Standard
SELCO	Solar Electric Light Company
SHS	solar home system
SREC	Solar Renewable Energy Credit
SSC	solar service center
TES	thermal energy storage
UNDP	United Nations Development Program
UNEP	United Nations Environment Program
USAID	United States Agency for International Development
VWU	Vietnam Women's Union
Wp	watt peak

BIBLIOGRAPHY

Begley, Ed, Jr. *Ed Begley, Jr.'s Guide to Sustainable Living: Learning to Conserve Resources and Manage an Eco-Conscious Life.* New York: Clarkson Potter, 2009.

Böer, Karl W., and Ester Riehl. *The Life of the Solar Pioneer Karl Wolfgang Böer.* New York: iUniverse, 2010.

Bradford, Travis. *Solar Revolution: The Economic Transformation of the Global Energy Industry.* Cambridge, Mass.: MIT Press, 2005.

Brown, Lester. *Plan B: 4.0 Mobilizing to Save Civilization.* New York: W. W. Norton, 2009.

Butti, Ken, and John Perlin. *A Golden Thread: 2500 Years of Solar Architecture and Technology.* New York: Van Nostrand Reinhold, 1980.

Chiras, Dan, with Robert Aram and Kurt Nelson. *Power from the Sun: A Practical Guide to Solar Electricity.* Gabriola Island, British Columbia: New Society Publishers, 2012.

Clarke, Arthur C. *How the World Was One: Beyond the Global Village.* New York: Bantam, 1992.

Clean Tech Nation: How the U.S. Can Lead in the New Global Economy. New York: HarperBusiness, 2012.

Climate Central. *Global Weirdness: Severe Storms, Deadly Heat Waves, Relentless Draughts, Rising Seas, and the Weather of the Future.* New York: Pantheon, 2012.

de Soto, Hernando. *The Mystery of Capital: Why Capitalism Triumphs in the West and Fails Everywhere Else.* New York: Basic Books, 2000.

Ewing, Rex, and Doug Pratt. *Got Sun? Go Solar.* Masonville, Colo.: Pixyjack Press, 2009.

Flannery, Tim. *The Weather Makers: How Man Is Changing the Climate and What It Means for Life on Earth*. New York: Atlantic Monthly Press, 2005.

Gelbspan, Ross. *Boiling Point*. New York: Basic Books, 2004.

———. *The Heat Is On*. Boston: Addison-Wesley, 1997.

Goodell, Jeff. *Big Coal: The Dirty Secret Behind America's Energy Future*. Boston: Mariner Books, 2007.

Goodstein, David. *Out of Gas: The End of the Age of Oil*. New York: W. W. Norton, 2004.

Gore, Al. *An Inconvenient Truth: The Planetary Emergence of Global Warming and What You Can Do About It*. New York: Rodale, 2006.

Hankins, Mark. *Stand-Alone Solar Electric Systems: The Earthscan Expert Handbook for Planning, Design and Installation*. London: Earthscan, 2011.

Hanncock, Graham. *Lords of Poverty: The Power, Prestige, and Corruption of the International Aid Business*. New York: Atlantic Monthly Press, 1989.

Hartmann, Thom. *The Last Hours of Ancient Sunlight*. New York: Three Rivers Press, 2004.

Hawken, Paul. *Blessed Unrest: How the Largest Social Movement in History Is Restoring Grace, Justice and Beauty to the World*. New York: Penguin Books, 2007.

———. *The Ecology of Commerce: A Declaration of Sustainability*. New York: HarperBusiness, 1993.

Heinberg, Richard. *Powerdown: Options and Actions for a Post-Carbon World*. Gabriola Island, British Columbia: New Society Publishers, 2004.

Inslee, Jay, and Bracken Hendricks. *Apollo's Fire: Igniting America's Clean Energy Economy*. Washington, D.C.: Island Press, 2008.

Jonnes, Jill. *Empires of Light: Edison, Tesla, Westinghouse, and the Race to Electrify the World*. New York: Random House, 2003.

Kennedy, Danny. *Rooftop Revolution: How Solar Power Can Save Our Economy—and Our Planet—from Dirty Energy*. San Francisco: Berrett-Koehler Publishers, 2012.

Klare, Michael. *The Race for What's Left: The Global Scramble for the World's Last Resources*. New York: Metropolitan Books, 2012.

Kunstler, James Howard. *The Long Emergency: Surviving the Converging Catastrophes of the Twenty-First Century.* New York: Atlantic Monthly Press, 2005.

Leggett, Jeremy. *The Carbon War: Dispatches from the End of the Oil Century.* New York: Penguin, 1999.

Lovelock, James. *The Revenge of Gaia: Earth's Climate Crisis and the Fate of Humanity.* New York: Penguin/Allen Lane, 2006.

Lovins, Amory. *Reinventing Fire: Bold Business Solutions for the New Energy Era.* White River Junction, Vt.: Chelsea Green Publishing, 2012.

Mann, Michael E. *The Hockey Stick and Climate Wars.* New York: Columbia University Press, 2012.

McKibben, Bill. *The Deep Economy: The Wealth of Communities and the Durable Future.* New York: Times Books, 2007.

McLuhan, Marshall. *Understanding Media: The Extensions of Man.* New York: McGraw Hill, 1964.

Palz, Wolfgang. *Power for the World: The Emergence of Electricity from the Sun.* Singapore: Pan Stanford Publishing, 2011.

Perlin, John. *From Space to Earth: The Story of Solar Electricity.* Cambridge, Mass.: Harvard University Press, 2002.

Rich, Bruce. *Mortgaging the Earth: The World Bank, Environmental Impoverishment, and the Crisis of Development.* Boston: Beacon Press, 1994.

Roberts, Paul. *The End of Oil: On the Edge of a Perilous World.* Boston: Houghton Mifflin, 2004.

Romm, Joseph. *Hell and High Water: Global Warming—the Solution and the Politics—and What We Should Do.* New York: William Morrow, 2007.

Ruppert, Michael C. *Crossing the Rubicon: The Decline of the American Empire at the End of the Age of Oil.* Gabriola Island, British Columbia: New Society Publishers, 2004.

Sandalow, David. *Freedom from Oil: How the Next President Can End the United States' Oil Addiction.* New York: McGraw Hill, 2008.

Scheer, Hermann. *The Energy Imperative: 100 Per Cent Renewable Energy Now.* London: Routledge, 2012.

———. *Energy Autonomy: The Economic, Social and Technological Case for Renewable Energy*. London: EarthScan, 2007.

———. *The Solar Economy: Renewable Energy for a Sustainable Global Future*. London: James & James, 2002.

———. *A Solar Manifesto*. London: Jones & James, 1994.

Schumacher, E. F. *Small Is Beautiful: Economics as If People Mattered*. New York: Harper and Row, 1973.

Solar Energy International. *Solar Electric Handbook*. Boston: Pearson, 2012.

Speth, James Gustave. *Red Sky at Morning: America and the Crisis of the Global Environment*. New Haven, Conn.: Yale University Press, 2004.

Steiner, Christopher. *$20 Per Gallon: How the Inevitable Rise in the Price of Gasoline Will Change Our Lives for the Better*. New York: Grand Central Publishing, 2009.

Tidwell, Mike. *The Ravaging Tide: Strange Weather, Future Katrinas, and the Coming Death of America's Coastal Cities*. New York: Simon & Schuster, 2006.

Turner, Chris. *The Leap: How to Survive and Thrive in the Sustainable Economy*. Toronto: Random House Canada, 2011.

ACKNOWLEDGMENTS

I first must thank Patricia Forkan, my life partner for twenty-five years, who stood behind my tilting-at-windmills efforts over the years and encouraged me to write my first book about solar energy. This book came about thanks to the great science fiction writer Ben Bova, former *Omni* magazine editor and author of more than 125 books, who encouraged me to update my original book with stories and information about what had transpired in the intervening seven years. He inspired me to keep writing about solar.

I am grateful to publishing veteran Tom Doherty for backing this project from the beginning. A special debt of gratitude is owed to Paul Stevens at Tor/Forge, and to the awesomely diligent copyeditor Angela Gibson.

Inaccuracies are my responsibility, but changing facts are not; the trajectory of solar power outpaces the efforts of writers and journalists to keep up with the story.

Many people have helped me or taught me along the road on my thirty-year solar journey. I can't name all to whom a debt of gratitude is owed. Here's a short list in no particular order:

Audrey McClellan, Chris and Judith Plant, Ingrid Witvoet, Dr. H. Harish Hande, Kamal Kapadia, M. R. Pai, Thomas Pulenkav, Hemalata Rao, K. M. Udupa, the late Visanthi Pai, Diljeet Titus, Akanksha Chaurey, B. R. Praabhakara, Priyantha Wijesooriya, Dr. A. T. Ariyaratne, Lalith Gunaratne, Lal Fernando, Susantha Pinto, Michael Northrop, Peter Riggs, Peter Yarrow, Tom Tatum, Paul Maycock, John Kuhns, Stephan Schmidheiny, Rolf Gerling, Jeremy Leggett, Christine Eibs-Singer, Phil Larocco, Bob Freling, Robert Shaw, Oliver Davidson, Deborah McGlauflin, Charlie Benoit, Peter Lowenthal,

Shawn Long, Tran Danh, Pete Peterson, Pham Hanh Sam, Elisabeth Stern, Erich Stoeckli, the late Sir Arthur C. Clarke, Anil Cabraal, Chris Flavin, Lester Brown, Jon Naar, Gerry Gallo, Steve Allen, Jon Davison, David Zucker, Petra Schweizer, the late Hermann Scheer, Lisa Frantzis, Richard Hansen, Joel Weingarten, Isabel Marks, Ross Gelbspan, Craig Van Note, Hermann Oberli, John Perlin, Mike Eckhart, Henrietta Fiennes, Scott Sklar, Dr. Peter Varadi, Eric Usher, Richard Bleiden, Julia Hamm, Jigar Shah, Monique Hanis, Mark Trexler, Daniel Kammen, Rhone Resch, Adam Friedensohn, John Thornton, Larry Kazmerski, Tom Leyden, Charlie Gay, Bernard McNellis, Dr. Peter Bourne, David Zucker, Michael Crossetti, Pete Myers, Bob Wallace, William Wallace, Joan Martin-Brown, Edward Goldsmith, Stanford Ovshinsky, Ed Begley, Jr., Khalid Shams, Mao Yinqi, Mathew Mendis, Mike Niklas, Dr. Surehsh Hurry, Charm Muchenje, Colonel Chhatra Gurung, Jagan Nath Shrestha, Wade Greene, Dak and Tek Bahadur Gurung, Tej Gauchan, Gibson Mandishona, Wolfgang Palz, Chaz Feinstein, Hazel Henderson, Tim Ball, Robin Broadfield, Jeff Leonard, Jack Vanderryn, Dave Hollister, Yug Tamrakar, Salyia Ranasinghe, Denis Hayes, NguTrinh Quang Dung, Larry Rockefeller, the late Harvey Forrest, Joel Weingarten, Bill Jordan, Mark Bent, Angelina Galiteva, Juan Rodriguez, Jorgdieter Anhalt, Mark Hankins, S. David Freeman, Larry Sherwood, Jeremy Rifkin, David Sandalow, Gery Critchley, Marlene Brown, the late Walt Ratterman, Johnny Weiss, Bill Gordon, Dell Jones, Gernot Oswald, Robert "Bud" Annan, Dipal Barua, John Naisbitt, Phil Coupe, Dan Weinman, Windy Dankoff, Brad Rose, Luis Reyes, Maureen McIntyre, My Hoa, the late Sir Kenneth Kleinwort, Tony Clifford, Lee Bristol, Titus Brenninkmeijer, Bob Thomason, Ellin Todd, Gerry Braun, Andrew Truitt, Katherine Brigham, Dr. Ron Cooper, Jan Jaremko, Brian Desmond, Roger Box, Jim Sharman, and Steve Lamb.

INDEX

ABOUT THE AUTHOR

Neville Williams is a former journalist, adventurer, activist, world traveler, and Vietnam War correspondent whose articles have appeared in *The New York Times Magazine, The New Republic, The Nation, Outside, New Times, Nature, Solar Today,* and other publications and newspapers. He began promoting solar energy as a media specialist at the U.S. Department of Energy during the Carter administration and later became interested in climate change and sustainable development while working as the national media director for Greenpeace in Washington, D.C. In 1990 he launched the nonprofit Solar Electric Light Fund to bring solar power to people without electricity in the developing world, and later started SELCO, which managed solar enterprises in India, Sri Lanka, and Vietnam. In 2005 he founded Standard Solar Inc. in Maryland to deliver solar power to residential and commercial customers in the United States. He lives in Naples, Florida. Website: *www.nevillewilliams.com.*